Recent Topics in Maintenance Management

Edited by Tamás Bányai

Published in London, United Kingdom

Recent Topics in Maintenance Management
http://dx.doi.org/10.5772/intechopen.1001757
Edited by Tamás Bányai

Contributors
Abdelmajid Farid, Amir Zare, Arameh Bisadi, Christian Okonta, Christopher Eboigbe, Hans Wernher van de Venn, Jaroslav Zajíček, Kiavash Fathi, Lars Magnus Hvattum, Mohamed El Khaili, Mohamed Rafik, Ralphael Edokpia, Redouane Fila, Tamás Bányai, Věra Pelantová

First published in London, United Kingdom, 2024 by IntechOpen
IntechOpen is the global imprint of INTECHOPEN LIMITED, registered in England and Wales, registration number: 11086078, 167-169 Great Portland Street, London, W1W 5PF, United Kingdom

British Library Cataloguing-in-Publication Data
A catalogue record for this book is available from the British Library

Additional hard and PDF copies can be obtained from orders@intechopen.com

Recent Topics in Maintenance Management
Edited by Tamás Bányai
p. cm.

This title is part of the Industrial Engineering and Management Book Series, Volume 6
Topic: Operational Excellence
Series Editor: Fausto Pedro Garcia Marquez
Topic Editor: Stuart So

Print ISBN 978-0-85466-917-2
Online ISBN 978-0-85466-916-5
eBook (PDF) ISBN 978-0-85466-918-9
ISSN 3029-0511

For EU product safety concerns:
IN TECH d.o.o., Prolaz Marije Krucifikse Kozulić 3, 51000 Rijeka, Croatia,
info@intechopen.com or visit our website at intechopen.com.

IntechOpen Book Series

Industrial Engineering and Management

Volume 6

Aims and Scope of the Series

Industrial Engineering and Management (IEM) is a discipline that focuses on optimizing complex processes and systems within various industries. It involves the integration of engineering, business, economics, mathematics, and behavioral sciences to improve efficiency, productivity, quality, and overall performance in organizations. Key aspects of Industrial Engineering and Management include: Process Optimization; System Analysis and Design; Quality Control and Management; Supply Chain Management; Operations Management; Human Factors and Ergonomics; Project Management; Cost Analysis and Financial Management; Decision Analysis.

Overall, Industrial Engineering and Management aims to optimize resources, improve processes, enhance productivity, and ensure the effective and efficient utilization of all elements involved in the production or delivery of goods and services. It is crucial in today's competitive business environment for organizations to stay efficient and competitive.

Production Engineering and Operational Excellence are fields of study and practices that focus on optimizing and improving the manufacturing and production processes within an organization. It combines principles from engineering, management, and operational strategies to enhance productivity, efficiency, quality, safety, and sustainability in the production of goods and services.

Here are the key components of Production Engineering and Operational Excellence: Process Optimization; Operational Excellence; Manufacturing Systems Design; Quality Management; Supply Chain Optimization; Production Planning and Scheduling; Automation and Technology Integration; Health, Safety, and Environmental Management; Cost Management; Performance Measurement and Key Performance Indicators (KPIs); Continuous Improvement and Innovation.

Production Engineering and Operational Excellence are crucial for organizations aiming to stay competitive in the global market by achieving high levels of efficiency, quality, and customer satisfaction while optimizing resources and minimizing waste. It is a multidisciplinary approach that encompasses engineering principles, management strategies, and the effective use of technology to drive operational success.

Meet the Series Editor

Fausto Pedro Garcia Marquez is a Full Professor at UCLM, Spain, with accreditation since 2013. He also holds the position of Honorary Senior Research Fellow at Birmingham University, UK, and serves as a Lecturer at the Postgraduate European Institute. In addition to these roles, Fausto has experience as a Senior Manager at Accenture from 2013 to 2014. He earned his European Ph.D. with the highest distinction. Throughout his career, Fausto has received numerous awards and honors. These include the Nominate Prize (2022), Gran Maestre (2022), Grand Prize (2021), Runner Prize (2020), and Advancement Prize (2018), as well as Runner (2015), Advancement (2013), and Silver (2012) by the International Society of Management Science and Engineering Management (ICMSEM). He was also the recipient of the First International Business Ideas Competition 2017 Award. Fausto's contributions extend to academic publishing, with over 242 papers to his name. Notably, his work has been recognized in journals like "Applied Energy" (Q1, IF 9.746, Best Paper 2020) and "Renewable Energy" (Q1, IF 8.001, Best Paper 2014). His affiliations include the editorial and authorship roles in more than 50 books, with publications through respected publishers such as Elsevier, Springer, Pearson, Mc-GrawHill, IntechOpen, IGI, Marcombo, and AlfaOmega. He has authored over 100 international chapters and holds 6 patents. Fausto serves as the Editor of 5 International Journals and is a Committee Member for more than 70 International Conferences. His research portfolio encompasses being the Principal Investigator in 4 European Projects, 8 National Projects, and participating in over 150 projects involving universities and companies. His areas of expertise and research interests span Artificial Intelligence, Maintenance, Management, Renewable Energy, Transport, Advanced Analytics, and Data Science. Fausto is a recognized Expert in the European Union in AI4People (EISMD) and ESF. He also serves as the Director of www.ingeniumgroup.eu, holds the status of Senior Member at IEEE since 2021, and has been honored as an Honorary Member of the Research Council of the Indian Institute of Finance since 2021. Fausto is also the Committee Chair of The International Society for Management Science and Engineering Management (ISMSEM) since 2020.

Meet the Volume Editor

Tamás Bányai received a master's degree in 1993 and a Ph.D. in 1999, both from the University of Miskolc, Hungary, where he is currently a full-time professor. He has 30 years of teaching and research experience in the design and control of materials handling systems and supply chain management, with special emphasis on heuristic optimization of large-scale systems. He has published more than 200 research papers, book chapters, and conference proceedings. He has been a member and manager of more than fifty national and international R&D projects. Away from academia, Prof. Bányai's other interests include playing the piano and taking photographs.

Contents

Preface

This book offers a selection of chapters on maintenance, and promoting new research results in the field. The book covers six topics, determined by the theoretical and practical aspects of maintenance.

Chapter 1, "Introductory Chapter: Trends of Maintenance in the Industry 4.0 Era", focuses on the application potentials of Industry 4.0 technologies in production and services to improve the efficiency of maintenance processes. In this chapter, the author discusses digital twin, augmented reality, heuristic and metaheuristic optimization, and machine learning because these tools, methods, and technologies can increase the efficiency of maintenance and transform conventional maintenance solutions into cyber-physical solutions through the digitalization and Internet of Things (IoT) technologies.

Chapter 2, "New Maintenance Management Topics", discusses novel trends in maintenance management focusing on characteristics of maintenance management and the status of maintenance issues in organizational practice. The chapter explains that the combination of knowledge in the field of maintenance and computer sciences has a great impact on human resources because it can significantly influence competence models in novel maintenance systems.

Chapter 3, "Maintenance Execution: What and How – A PDCA Approach", discusses the how-to-do aspect of maintenance execution involving the capability and ability of the shopfloor technicians. By adopting the Plan-Do-Check-Act (PDCA) methodology, a systematic approach to maintenance execution is proposed that focuses on both planning and implementation of maintenance from a reliability point of view.

Chapter 4, "Optimizing Spare Part Management for Vessels in Liner Shipping", proposes a novel methodology to optimize inventory management of spare parts required for maintenance of seagoing vessels. The approach demonstrates the problem of inventory both in warehouses and on vessels. The chapter considers liner shipping. In the model, each vessel follows a pre-planned itinerary and several preventive maintenance tasks have been scheduled over time.

Chapter 5, "Data, Models, and Performance: A Comprehensive Guide to Predictive Maintenance in Industrial Settings", discusses the impact of data, data quality, model maintenance, and model interpretability on the performance and acceptability of predictive maintenance. The chapter describes the importance of digitalization in the case of predictive maintenance in original equipment manufacturers (OEMs) and small and medium-sized enterprises (SMEs).

Chapter 6, "Contribution of Artificial Intelligence to Industrial Maintenance in the Field of Mechanics", highlights the importance of artificial intelligence (AI) in industrial maintenance. Global industry uses both preventive and corrective maintenance strategies, but in the case of a dynamic environment, proactive aspects become more and more important; therefore, the chapter focuses on proactive maintenance involving monitoring and verification of root causes.

The aim of this book is to help students as well as managers and researchers to understand and appreciate the concept, design, and implementation of maintenance management. The editors thank the chapter authors for their scientific contributions. The chapters were edited and published following a rigorous selection process. We also wish to thank and acknowledge the many individuals who helped us throughout the editorial process that made this book possible.

Tamás Bányai
Institute of Logistics,
University of Miskolc,
Miskolc, Hungary

Chapter 1

Introductory Chapter: Trends of Maintenance in the Industry 4.0 Era

Tamás Bányai

1. Introduction

In order to increase the efficiency of production and service activities, companies need to pay increasing attention to ensuring the functionality of logistics and technological resources of systems and processes, which can be effectively supported by a well-functioning maintenance solution. In the field of planning and management of maintenance activities, a number of typical research directions can be identified that contribute significantly to improving the efficiency and reliability of production and service activities. The literature discusses a wide range of maintenance policies, such as predictive maintenance, corrective maintenance, preventive maintenance, condition-based maintenance, emergency maintenance, risk-based maintenance, planned maintenance,

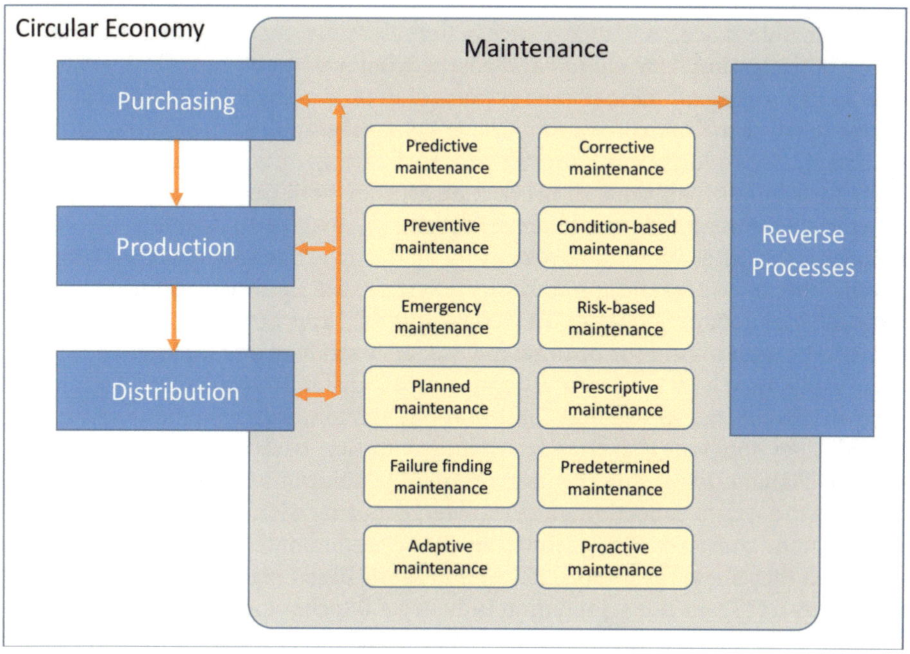

Figure 1.
Maintenance in the value chain.

prescriptive maintenance, failure finding maintenance, predetermined maintenance, adaptive maintenance, total productive maintenance, or proactive maintenance. The various types of maintenance strategies have a great impact not only on the core processes of the value chain including the production processes but also significantly influence the efficiency of connected purchasing, distribution, and inverse or reverse processes (see **Figure 1**).

This book provides the reader with an overview of the latest trends in maintenance activities to improve the efficiency of production and service systems. As the chapters illustrate, maintenance activities can improve the efficiency of production activities in a way that not only increases the level of customer service but also the sustainability of systems. Appropriate maintenance strategies are also becoming increasingly important in supply chains, as the proper performance of maintenance activities in road, rail, water, and air transport can make a major contribution to improving the reliability of global supply chains. The chapters in this book will help you understand these potential benefits.

2. Industry 4.0 technologies in maintenance

Jardine et al. [1] show in their research focusing on machinery diagnostics and prognostics implementing condition-based maintenance that the operation of condition-based maintenance is generally based on data acquisition, data processing, and maintenance decision-making, where the data acquisition and data processing are significantly influenced by diagnostics and forecasting of resources' status. This research also highlights the importance of up-to-date sensor systems, multi-sensor solutions, which can lead to big data problems to be solved. Maintenance policy can also significantly influence the availability, flexibility, cost efficiency, and performance of maintenance, as Bányai shows in her researches focusing on maintenance strategy optimization from energy and cost efficiency point of view [2, 3]. Wang also discusses in a survey of maintenance policies of deteriorating systems [4] that there is a wide range of maintenance strategies, which can be applied depending on the objective of the production and service process focusing on age replacement, periodicity, failure limit, repair time, or prevention aspects. Imperfection is a very important characteristics of maintenance. Pham and Wang [5] analyze 40 different mathematical models of maintenance policies focusing on the imperfection, and they summarize that reliability measures and the state of the systems are significantly influenced by the chosen maintenance strategy. Therefore, suitable, appropriate optimization models have to be used to find the optimal maintenance strategy and maintenance policy. Ahmad and Kamaruddin also discuss and compare different maintenance strategies in an overview of time-based and condition-based maintenance in industrial applications [6]. They conclude that time-based and condition-based maintenance have their own characteristics, and the integration of IT solutions (data determination, data acquisition, data collection, data transfer, data analysis, and decision-making) depends on the chosen strategy. However, time-based maintenance policy is widely known, but they show the importance of condition-based maintenance, which needs state-of-the-art IT solutions, including Industry 4.0 technologies.

Machine learning and the application of artificial intelligence shows significant potential in improving the performance of maintenance solutions. Carvalho et al. discusses in their research [7] the application of machine learning technologies in predictive maintenance and shows that Linear Regression, Gaussian Process regression,

Bayesian Network, Support Vector Machine, Multi-Gene Genetic Programming, Artificial Neural Network, Convolution Neural Network, Long Short-Term Memory Network, Deep Learning, Hierarchical Clustering, Gradient Boost, and k-means are the most widely used machine learning methods in predictive maintenance. These machine learning methods can significantly improve the design and operation of maintenance solutions, especially in the case of robust, large-scale industrial applications. As a practical scenario shows, Analytic Hierarchy Process is a very appropriate methodology to choose the optimal maintenance strategy because maintenance strategy or maintenance policy selection can be described as a complex, multi-tier decision-making problem. As the research by Bevilacqua and Braglia show [8], Analytic Hierarchy Process can be used to arrange the characteristics and parameters of maintenance strategies and policies, and the comparison of different maintenance policies can be based on the pairwise comparison of judgment aspects. The production and service processes can be described either as deterministic or as stochastic systems. In the case of maintenance, because of the uncertainties of resources and the environment, the decision-making models are generally uncertain models, and the systems can be described as stochastically deteriorating systems, where inspection frequency and maintenance degree significantly influence the efficiency of the chosen maintenance policy, as shown by Alaswad and Xiang [9]. As Cho and Parlar describe in their research [10], the availability, efficiency, sustainability, and flexibility of maintenance are significantly influenced by a wide range of subprocesses and sub models to be optimized, including repair models, inventory models, replacement models, and inspection models. The performance of maintenance operations and the maintenance policy depends on both technological and logistics aspects, where logistics aspects include not only inventory management problems but also purchasing, distribution, and transportation problems.

As a research by Peng et al. shows [11], machine prognostics is a core problem of maintenance (especially in the case of condition-based maintenance). They suggest a novel systematic maintenance decision framework, which integrates a wide range of Industry 4.0 technologies, including feature selection based on Principal Component Analysis, Genetic Algorithm, and Support Vector Machine; data training; diagnostics, forecasting, and prognostics based on real-time sensor data; and maintenance schedule influenced by the precision of prognostics and maintenance cost function.

Augmented reality is also an important Industry 4.0 technology, which can be integrated into digital twin solutions and can play an important role in the education and training of maintenance and support maintenance operations by external experts. Palmarini et al. [12] summarize the application potentials of augmented reality in maintenance, and their research results indicate a high fragmentation of hardware, software, and solutions. This fragmentation has a great impact on the complexity of augmented reality solutions. Augmented reality solutions lead to the integration of emerging technologies and result in the novel tele maintenance paradigm. The prognostic and the forecasting play an important role in the organization of maintenance operations, therefore it is important to have up-to-date solutions to decrease of error value in forecasting and prognostic processes. As Lee et al. show in research focusing on intelligent prognostics tools and e-maintenance [13], the integration of different Industry 4.0 technologies and Internet of Things solutions leads to the transformation of conventional maintenance strategies and policies (e.g. fail and fix policy) into e-maintenance policies (e.g. predict and prevent). Swanson also discusses this topic and concludes that fire-fighting strategies will be replaced by proactive maintenance policies including preventive, predictive, and total productive maintenance [14]. Garga and Deshmukh concluded [15] that maintenance management can be represented as an integrated approach because it includes a wide range of optimization models and methods, various maintenance techniques, different scheduling algorithms and approaches, telecommunication, and IT solutions. As Muller et al. conclude [16], this integration has important aspects from a collaborative environment, pertinent knowledge, and intelligence point of view.

As the above-mentioned literature review shows, Industry 4.0 technologies play an important role in the development, improvement, and operation of maintenance, and they can significantly improve efficiency. As discussed in previous studies [17], Industry 4.0 technologies are increasingly bringing to the fore advanced maintenance strategies that enable real-time optimization of maintenance. This is based on the idea that a logical layer, which is a digital twin of the real-world processes, can be created from the physical layer represented by the resources to be maintained by means of a suitable sensor network. The sensor network naturally generates such a large amount of data from the real system that it requires the use of advanced big data technologies to process it. From the digital twin, a real-time simulation model can be generated, in which real-time simulations can be performed using a suitable simulation software with a model of the real system with current parameters, and the future state of the system and the resources that build the system can be predicted with high accuracy. Based on these predictions, real-time optimization can be done to define a maintenance strategy to perform the maintenance activities required by the predictions in a cost-effective manner.

The digital twin retrieves information from a centralized database, which is linked to the Enterprise Resource Planning (ERP) and Manufacturing Execution System (MES). The optimization results of the real-time simulation and forecasting are linked to both the digital twin, the real-world system, the ERP, and the MES (see **Figure 2**).

The Industry 4.0 maturity of companies also influences the success of digitization of conventional maintenance because the available hardware and software components, production, logistics, and business processes have a great impact on the integration of Internet of Things solutions and the transformation of conventional maintenance operations into e-maintenance and maintenance in a cyber-physical environment.

The optimization of maintenance-related design and operation problems can be characterized as NP-hard optimization problems. Veres et al. [18] conclude in their

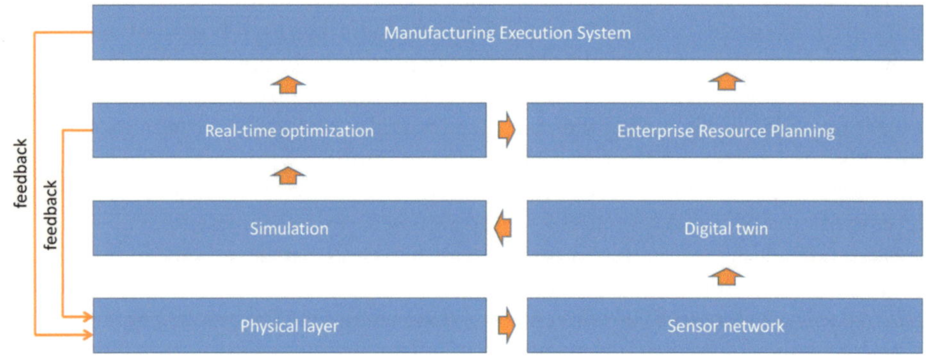

Figure 2.
Concept of real-time optimization of maintenance policy.

research work that the design and operation tasks of industrial environments are determined by a wide range and number of deterministic and stochastic parameters. This increased number of variables and parameters results in enormous computational time for the exact solution, and this problem highlights the importance of heuristic and metaheuristic algorithms, which are useful to solve NP-hard optimization problems, as shown in the case of different maintenance-related problems:

- differential evolution algorithm is used by Feng et al. [19] to optimize a selective maintenance problem with stochastic durations,
- an adaptive large neighborhood search approach was proposed by Liu et al. [20] for the optimization of maintenance routing and scheduling,
- Arzanlou and Sardroud apply Particle Swarm Optimization for budget allocation and scheduling of maintenance operations [21],
- Zhang et al. [22] propose an improved Ant Colony Optimization for the operational aircraft maintenance routing problem.

As the above-described heuristic and metaheuristic approaches show, maintenance optimization is a complex mathematical problem where new models and methods are available to improve the performance of design and operation tasks.

3. Conclusions

Manufacturing and service companies must make increasing efforts to meet the dynamically changing needs of their customers in a cost-effective, timely, and high-quality way. This requires a production and service infrastructure that operates reliably and continuously. An important prerequisite for this reliable operation is an effective maintenance strategy to ensure that production and service systems operate efficiently. In this chapter, the author presented the main Industry 4.0 technologies (focusing on digital twin, augmented reality, heuristic and metaheuristic optimization, and machine learning) that can increase the efficiency of maintenance activities and transform conventional maintenance into maintenance in a cyber-physical environment through the integration of digitalization and Internet of Things technologies.

Conflict of interest

The authors declare no conflict of interest.

Author details

Tamás Bányai
Institute of Logistics, University of Miskolc, Miskolc, Hungary

*Address all correspondence to: tamas.banyai@uni-miskolc.hu

References

[1] Jardine AKS, Lin D, Banjevic D. A review on machinery diagnostics and prognostics implementing condition-based maintenance. Mechanical Systems and Signal Processing. 2006;**20**:1483-1510. DOI: 10.1016/j.ymssp.2005.09.012

[2] Bányai Á. Impact of agile, condition-based maintenance strategy on cost efficiency of production systems. Advances in Production Engineering & Management. 2023;**18**:317-326. DOI: 10.14743/apem2023.3.475

[3] Bányai Á. Energy consumption-based maintenance policy optimization. Energies. 2021;**14**:5674. DOI: 10.3390/en14185674

[4] Wang H. A survey of maintenance policies of deteriorating systems. European Journal of Operational Research. 2002;**139**:469-489. DOI: 10.1016/S0377-2217(01)00197-7

[5] Pham H, Wang H. Imperfect maintenance. European Journal of Operational Research. 1996;**94**:425-438. DOI: 10.1016/S0377-2217(96)00099-9

[6] Ahmad R, Kamaruddin S. An overview of time-based and condition-based maintenance in industrial application. Computers and Industrial Engineering. 2012;**63**:135-149. DOI: 10.1016/j.cie.2012.02.002

[7] Carvalho T, Soares F, Vita R, Francisco R, Basto J. A systematic literature review of machine learning methods applied to predictive maintenance. Computers and Industrial Engineering. 2019;**137**:106024. DOI: 10.1016/j.cie.2019.106024

[8] Bevilacqua A, Braglia B. Analytic hierarchy process applied to maintenance strategy selection. Reliability Engineering and System Safety. 2000;**70**:71-83. DOI: 10.1016/S0951-8320(00)00047-8

[9] Alaswad S, Xiang Y. A review on condition-based maintenance optimization models for stochastically deteriorating system. Reliability Engineering and System Safety. 2017;**157**:54-63. DOI: 10.1016/j.ress.2016.08.009

[10] Cho D, Parlar M. A survey of maintenance models for multi-unit systems. European Journal of Operational Research. 1991;**51**:1-23. DOI: 10.1016/0377-2217(91)90141-H

[11] Peng Y, Dong M, Zuo MJ. Current status of machine prognostics in condition-based maintenance: A review. International Journal of Advanced Manufacturing Technology. 2010;**50**:297-313. DOI: 10.1007/s00170-009-2482-0

[12] Palmarini R, Erkoyuncu JA, Roy R, Torabmostaedi H. A systematic review of augmented reality applications in maintenance. Robotics and Computer-Integrated Manufacturing. 2018;**49**:215-228. DOI: 10.1016/j.rcim.2017.06.002

[13] Lee J, Ni J, Djurdjanovic D, Qiu H, Liao H. Intelligent prognostics tools and e-maintenance. Computers in Industry. 2006;**57**:476-489. DOI: 10.1016/j.compind.2006.02.014

[14] Swanson L. Linking maintenance strategies to performance. International Journal of Production Economics. 2001;**70**:237-244. DOI: 10.1016/S0925-5273(00)00067-0

[15] Garg A, Deshmukh A. Maintenance management: Literature review and

directions. Journal of Quality in Maintenance Engineering. 2006;**12**:205-238. DOI: 10.1108/13552510610685075

[16] Muller A, Crespo MA, Iung B. On the concept of e-maintenance: Review and current research. Reliability Engineering and System Safety. 2008;**93**:1165-1187. DOI: 10.1016/j.ress.2007.08.006

[17] Bányai Á, Bányai T. Real-time maintenance policy optimization in manufacturing systems: An energy efficiency and emission-based approach. Sustainability. 2022;**14**:10725. DOI: 10.3390/su141710725

[18] Veres P, Illés B, Landschützer C. Supply chain optimization in automotive industry: A comparative analysis of evolutionary and swarming heuristics. Lecture Notes in Mechanical Engineering. 2018;**1**:666-676. DOI: 10.1007/978-3-319-75677-6_57

[19] Feng X, Chen X, Zhang L, An Y. Optimization of selective maintenance problem with stochastic durations in mission-oriented system subjecting to s-dependent competing risks. Expert Systems with Applications. 2024;**245**:123068. DOI: 10.1016/j.eswa.2023.123068

[20] Liu L, Xiao Y, Yang J. Daily optimization of maintenance routing and scheduling in a large-scale photovoltaic power plant with time-varying output power. Applied Energy. 2024;**360**:122793. DOI: 10.1016/j.apenergy.2024.122793

[21] Arzanlou A, Sardroud JM. Enhancing project performance: Particle swarm optimization for optimal budget allocation and maintenance scheduling. KSCE Journal of Civil Engineering. 2024;**28**:1635-1644. DOI: 10.1007/s12205-024-1348-1

[22] Zhang Q, Chan FTS, Fu X. Improved ant Colony optimization for the operational aircraft maintenance routing problem with cruise speed control. Journal of Advanced Transportation. 2023;**2023**:8390619. DOI: 10.1155/2023/8390619

Chapter 2

New Maintenance Management Topics

Věra Pelantová and Jaroslav Zajíček

Abstract

This chapter deals with new topics in maintenance management. The need for maintenance as a result of changes in the substantial environment of organisations increases. Based on current maintenance problems in organisations and social and environmental needs of society, key management trends can be deduced through the system analysis. It follows a large area of quite changing legislation. The field of Artificial Intelligence and the Internet of Things and so on also come into play in maintenance. The chapter is also based on the practice of authors in this field. It therefore affects the area of production equipment, human resources, software support, costs and the material base. Maintenance management risks are also significant. Without clear trends, organisations cannot direct their strategy and thereby effectively manage their own maintenance. This chapter is intended to help organisations strengthen their overall competitiveness through maintenance management.

Keywords: maintenance management, maintenance, trends, problems, organisation, competitiveness, object

1. Introduction

The economic maturity and the current situation of a given country determine the state of maintenance management and quality management in organisations. The dynamics of the development of maintenance management can be documented in the example of the Czech Republic. As a result of changes in the essential environment of organisations, the need for complex maintenance (asset management) is increasing. At first sight, the economic field does not present this trend. However, higher maintenance costs are evident. In particular, the increase in taxes in the Czech Republic and two higher value-added tax rates are essential. The price of the highway mark is higher. Changes characterised by increases in social and health insurance for employees and self-employed persons, such as maintenance employees, affect not only the labour costs of maintenance prices. To this must be added higher travel allowances. All this means an increase in the cost of logistics services and labour, including maintenance [1]. In this context, publication [2] calls for the inclusion of crisis management in organisations and the use of a process perspective. Emphasis is to be placed, among other things, on avoiding waste of resources and energy.

The legislative area is also an essential environment, as in the article [3]. Legislative changes in the Czech Republic mainly concern the strengthening of fire safety and the environment towards a circular economy in relation to industrial activities. The assessment of compliance with general requirements for the competence of providers and the technical requirements for certain materials such as paints, sealants, leather and textiles are also essential. The aim is to reduce the production of microplastics and microparticles in general. An open environment that will be yet legislated to a greater extent is modern information technology (including Industry 4.0 and so on).

This chapter provides a basic literature review of current research in the field of maintenance management. It presents a list of methods for maintenance management and lists its problem areas. It proposes a set of characters for the maintenance process, describes the current state of this issue in organisational practice and sets trends for the possible development of maintenance management in particular for the need of organisation's strategy.

2. Maintenance management in publications

The thematic direction of maintenance management must be conceived from the external context of organisational practice and from the theoretical results of maintenance research. Current publications on the field discuss the following areas.

There are corrective, preventive, and predictive maintenance as it is common in current organisational practice, the publication [4] wrote not only. When predictive maintenance is applied, a 46% saving in maintenance costs is based on this publication. However, it is necessary to increase the proportion of preventive maintenance compared to post-failure maintenance to reflect the cost reduction, as added in publication [5]. The maintenance of a multi-component system under uncertain operating conditions is discussed in the text [6]. Quite often, decisions are made regarding extending the lifetime of components that should already be replaced due to their wear and tear, as added by the text [5]. The publication [7] presents a custom metric oriented towards decision-making and forecasting in relation to predictive maintenance. The useful life of degrading components is determined.

The article [4] describes maintenance planning using the CBM (condition-based maintenance) method, here using aircraft maintenance as an example. It updates the list of maintenance tasks on a daily basis and tries to match them with the available hardware/software interface and resources. The scheduling is two-phase for equipment and for components, as for example in text [5]. At the same time, this paper concludes that a more complex system will fail faster due to the enormous number of components. A CBM methodology based on a combination of reinforcement learning and machine learning and remaining component lifetime is presented in the article [8]. Again, the goal is mainly to minimise the maintenance cost. Commonly, somewhat simplified component deterioration conditions are considered, which makes subsequent maintenance prediction difficult. For example, the impact of maintenance limitations on train availability, caused by various events of external context, is presented in the text [9]. It follows that it is clearly worthwhile to perform at least basic maintenance on objects. Not to omit maintenance altogether because of cost or other complications. The paper [10] investigates algorithms for the problem of facility maintenance frequency and moves to the next facility in similarities with the BGT method. The paper [11] addresses redundancy allocation for a series-parallel system. In doing so, each component can be characterised by two function states and further by its performance.

Maintenance scheduling is now supported using networks that also need to be topologically characterised, due to the reduction of maintenance overhead, which is addressed in the publication [12]. The text [13] describes data management issues for AI (Artificial Intelligence) and IOT (Internet of Things). It is necessary to ensure the reliability of data when the data are large. In addition, skilled personnel with different knowledge and skills than before are needed for maintenance in conjunction with AI and IOT. All of these lead to more accurate maintenance decisions. The study [14] describes the relationship of maintenance in the Industry 4.0 with respect to employee outcomes and develops a competency model of maintenance for the Industry 4.0, which in turn leads to the satisfaction of these employees. This is due to the change in maintenance employees' activities as well as material and information flows as a result of the introduction of the Industry 4.0 as a technological innovation and as a result of socio-economic changes. This model is linked to a hierarchical structure. The optimization strategy of predictive maintenance is presented in Ref. [15]. Digital twins of each object (i.e., an electronic copy of the object including the associated functions and properties) are created, as in the text [16], and their behaviour over time is monitored based on historical maintenance data. The goal here is to gain predictability of maintenance to avoid incurring unpredictable costs. The combination of spare parts management (management, circulation, performance, etc.) for maintenance and the Industry 4.0 is discussed in the article [17]. The connection between TPM (total productivity maintenance) and the Industry 4.0 is described in the research [18]. It establishes characters (parameters) of system sustainability also here on a hierarchical structure. It sees knowledge of TPM issues, employee involvement and support from the organisation's leadership as key success factors. The application of AI and IOT is now perceived by domestic society as favourable. However, the author of [19] points out that the power of words from an emotional perspective is not yet included in AI. Facility management in conjunction with IOT is described in the article [20]. It points out the lack of educated staff in this context, as does the text [16]. The software applied has a fundamental influence on the quality of the system implementation. In maintenance, the shortcomings of IOT in the form of not taking out the rubbish, not cleaning the corners and not cleaning the work area are evident. On the other hand, there is an effort to conceptualise modern facility management in a responsible way, with emphasis on social aspects (application of less employable people) and ecological aspects (application of eco-friendly means, recycling, and carbon footprint monitoring) and the corresponding application of diagnostics. A hot new feature is the inclusion of non-financial reporting in maintenance management, which means tracking characters that mainly describe social, environmental and safety areas. The text [16] mentions the problem of expanding the digitalisation of organisations in the persistent shortage of chips and therefore of object controllers. On the other hand, it sees a great benefit in more accurate corrections in production and maintenance and in the automatic adjustment of objects. The paper [21] then presents possibilities of data collection and value monitoring. The speed of data processing and the production of specific alerts about the state of the components in the system is high and certainly pleasing for users in maintenance. The open connectivity, interface options and open-source applications are advantageous. However, it does not recommend an anonymous cloud environment or linkage to a single brand to store and process data.

Resource	Method	Research	Main result	Key problem
[4]	CBM	Aircraft	Planning, cost reduction	Quality maintenance data
[5]	Markov processes	Component systems	Planning model for component system	Component life extension decision-making algorithm
[6]	Weibull distribution, stochastic programming	Component systems	Decision model in events	Problem analysis in different scenarios
[12]	Deep Q-Learning, Net topology	Network systems	Reduced overhead for network	Different failure scenarios
[8]	CBM, learning, Kaplan Meier Product Limit	System maintenance	Optimization to the CBM	The function of the component deteriorates unevenly
[9]	KPI	Trains	Simulation of impact of constraints	The model of the failing device
[14]	Publication review, interview	Publications and organisations	Competence model for Industry 4.0	Implementation of model into the organisation
[13]	Publication review, predictive analyses	Industry engineering + AI + IOT	Data management and AI + IOT system	The relationship of spare parts and Industry 4.0
[23]	Publication review	Application of machine learning	Maturity levels for learning	Object failure scenarios and its safety
[22]	Markov processes, machine learning	Industry organisations	Deep learning in systems and safety	Infrequent object failure scenarios and its safety
[28]	Publication review, audits	Audit documentation, organisation	Audits do not consider climate	Specification of risks of object during the audit
[24]	Discussion, Publication review	Wind power plants	Failures, sustainability	Premature object failure and costs
[18]	Publication review, expert interview	Industry organisations	TPM, Industry 4.0, sustainability	Impact of TPM on the sustainability
[27]	HAZOP, LOPA, Aaalysis of scenarios	Nuclear equipment	Risk assessment and safety	Failure scenario and risk analysis

Table 1.
The comparison table of survey results in several mentioned publications in relation to the issue of maintenance management (own of authors by [4–37]).

Deep learning (a type of machine learning algorithm development for artificial neural networks with extensive features) in predictive-maintained critical safety systems is described in the text [22]. The study [23] discusses the maturity levels of a machine learning system (basic device learning from experience without intentional programming by another person) for predictive maintenance. The characters of the system are object reliability, performance and intelligence. However, it encounters imprecise machine learning terminology and poor access to the necessary maintenance data in the organisation. A major problem is the cost of maintenance and the difficulty of integrating the systems overall, both technically and organisationally. Sustainability in this area is also under scrutiny. Kinds of turbine failures and maintenance with a focus on sustainability are presented in the text [24]. The key objects in this type of system are the bearings and their tribology. Hence, the maintenance costs of the system are important to know. It is also necessary to monitor the conditions under which nonconformities of sub-objects occur.

Donor funding and its effects on system maintenance, using water supply as an example, are explored in the text [25]. It identifies a number of problems in this context, such as expensive spare parts, delayed delivery of parts, imposition of foreign technology and disregard for local maintenance staff, which leads to complications in maintenance and its resulting cost to the end customer. The text [26] provides a discussion of moisture management in the building with respect to maintenance, which is becoming a major problem. The reference [27] focuses on avoiding employee exposure to ionising radiation during maintenance of a nuclear facility in this example. It assesses the risks of components. Again, it targets sustainability and system safety.

The issue of maintenance auditing and its limitations is addressed in the article [28]. The internal climate of the organisation, the influence of management power on maintenance productivity, psychosocial factors and so on are not included in audits. Auditors do not observe everything and often do not perceive serious problems and do not specify these findings in reports. The organisation then has no substantial basis for planning and management, not just maintenance.

A summary of survey results from several mentioned publications in relation to the issue of maintenance management is given in **Table 1**.

3. Methods for the maintenance

In relation to publications cited here and to the findings of the authors of this paper, it can be noted that methods used for planning and managing maintenance are varied. The combination of the time to failure of an object and its maintenance costs is frequent and other mathematical and statistical calculations of the reliability of objects or the determination of maintenance KPI (key productivity indicator), which follow more the economic side of the matter, as well as OEE (overall equipment effectiveness) are also applied.

More sophisticated methods include the Monte Carlo method, Markov processes, the Kaplan Meier Product Limit, the stochastic programming to the Deep Q-learning, which may encounter a lack of quality maintenance data. Due to the complexity of the maintenance system and its essential environment, in order to consider multiple factors and features, the multicriteria decision-making and the full consistently method are applied, and from the point of view of reliability and human safety, HAZOP (hazard and operability study) and LOPA (layer and protection analysis) methods are applied.

However, research often encounters imperfections in maintenance models and therefore many research works are based on a literature search and guided interviews with rank-and-file employees or experts in the maintenance field. Somewhat unique, but essential for facility maintenance planning, is the so-called the bamboo method (BGT).

Methods for the maintenance are more categorised in **Figure 1**. Methods are divided into basic and expanding type, quantitative and qualitative type, inductive

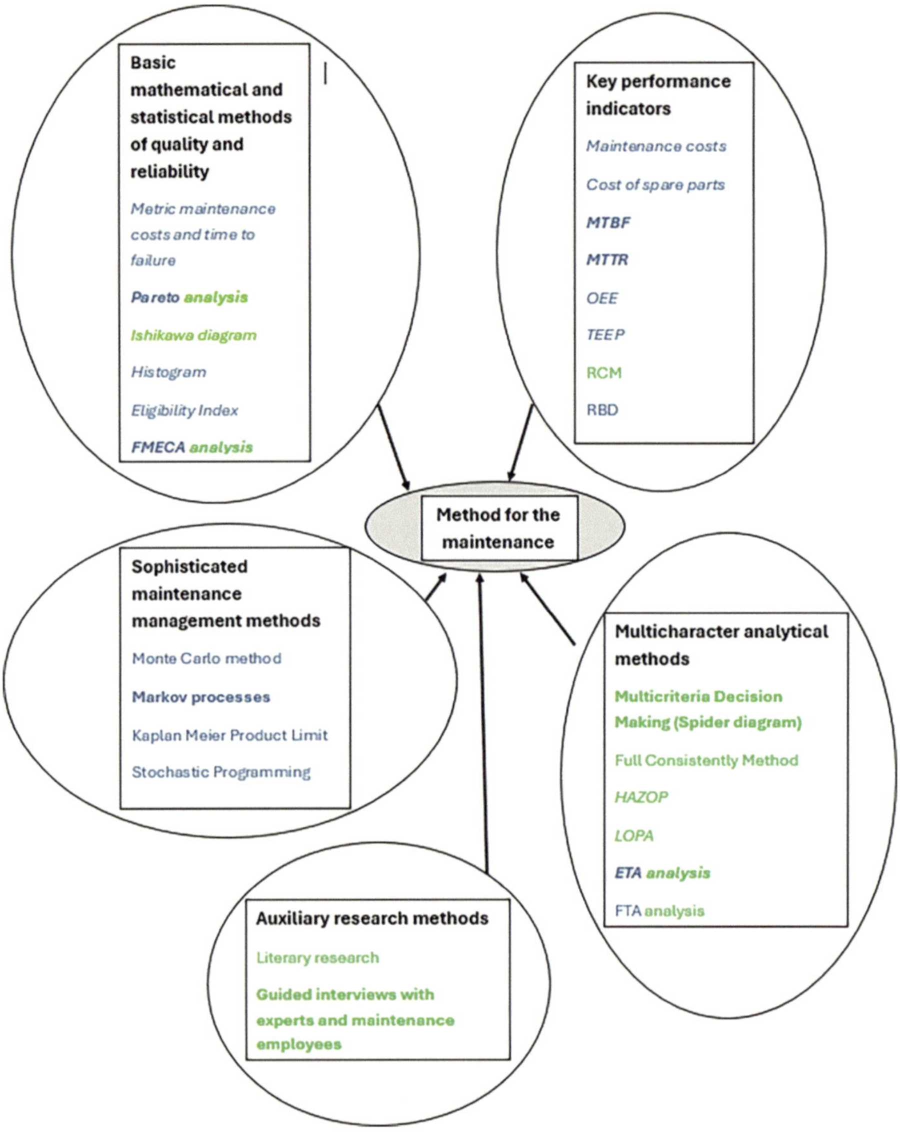

Figure 1.
Methods for the maintenance (own of authors by [30, 31, 36, 37]). Blue—Quantitative methods; Green—Qualitative methods; Bold—Basic reliability analysis methods; Italics—Inductive methods (bottom-up state analysis); No italics—Predominantly deductive method; MTBF—Mean time between failure; MTTR—Mean time to repair; TEEP—Total equipment effectiveness performance; RBD—Reliability block diagram; FMECA—Failure modes, effects and criticality analysis; FTA—Fault tree analysis; ETA—Event tree analysis.

and deductive type and according to the difficulty and the area of targeting. For example: Sophisticated maintenance management methods require valuable input data, are computationally demanding, are more expensive to apply and are therefore less used by organisations. Deductive methods help to explain the consequences of failure states.

Note: Two-colour marked methods can be of either type depending on the given data.

4. Problems in the maintenance management

The following problem areas arise from the above.

Essential for the prediction of maintenance is its appropriate planning and optimisation, for example, according to the text [4]. However, this is preceded by the collection and analysis of good-quality maintenance data. The component function deteriorates unevenly, for example, according to the reference [8].

The aim of surveys not only in these publications is usually to determine and reduce maintenance costs, as stated in the text [4]. Prediction methods from models need to be compared and evaluated. It is also necessary to analyse the problems in different maintenance scenarios according to the text [6], and to consider different failure scenarios, as added by the paper [12]. The effect of the inventory model on the maintenance system and its performance metrics needs to be investigated. Furthermore, a decision algorithm needs to be developed for extending the lifetime of components that should have already been replaced in the system, as reported, for example, in the article [5]. However, the safety aspect is not given enough priority in this direction.

Better computational procedures need to be developed for complex maintenance scenarios and a more heuristic approach needs to be incorporated into the scalability problem. More efficient approaches for evaluating maintenance planning that is intertwined with the use of networks are also needed. There is also a need to develop a model of a failing device, for example, a car, as most models are based on an ideal fully functional system, as the text [9] adds. There are also frequent oversights and existing difficulties in implementing maintenance models in organisations, as described, for example, in the text [14]. Even in simpler cases of BGT problems, not all variables (such as approximation ratios for similar objects) are known, which makes maintenance planning inaccurate, for example, according to the text [10]. Efficient data management within AI and IOT and appropriately skilled maintenance staff and a new business model for the organisation should lead to an efficient system, for example, text [13] believes. Furthermore, there is still little knowledge about the relationship between spare parts management and Industry 4.0. The research [18] also recommends investigating the impact of TPM on sustainability in different sectors of the economy. The publication [27] recommends improving the safety and sustainability of equipment through failure scenario analyses and risk analyses.

Research should focus on predictive maintenance with respect to safety-critical problems and low-frequency facility failure scenarios, which are often forgotten so far, as confirmed by publications [22, 23]. Also, the specification of facility risks should be addressed when the auditing of the maintenance process is performed as concluded by the article [28]. The question is also how to involve donation and aid policy in the maintenance, as pointed out in the text [25].

5. Determination of maintenance trends

Overall, it can be concluded that there is not yet a set of trends that should guide future research on the maintenance management. Therefore, this matter is further addressed. Based on the previous research [4–29] and according to the authors' own knowledge of maintenance management, 6 categories of problems in this area were determined using the Affinity diagram (in the book [36]). These are:

a. Component life extension (mentioned six times)

b. Methods in maintenance (prognostics, scenarios, audits and heuristics) (mentioned four times)

c. Maintenance personnel and their competencies (mentioned three times)

d. Spare parts (mentioned two times)

e. Sustainability (mentioned two times)

f. Artificial Intelligence (mentioned one time)

Categories (e) and (f) are still developing. Category (e) is linked to the environmental management system (according to the standard ISO 14001 [33]). The system sustainability attributes from the publication [18] are qualitative but harder to understand and to compare. Therefore, they need to be redefined. From the authors' experience, characters such as: system downtime (minutes), data flow rate (Mbit/s), number of conflicts and number of security incidents, can be recommended. Category (f) is still a kind of loose discipline, but it carries several risks and will have to be guarded in the future. The terminology is not professionally finalised according to standards ISO 9001 [32], EN 13306 [34] and ISO/IEC 27000 [35] and so on. The connectivity is not stable, although technically and software-wise the problem is being solved.

Category (a) is currently the most mentioned. It reflects the current situation with shortages of raw materials and spare parts and poor customer-supplier relations. The safety criterion is essential and only then should the cost item be addressed. Uncertainty is associated with category (b). Complex scenarios need to be dealt with in a predictive and auditable manner to obtain more realistic pictures of the system. This places increased demands on maintenance personnel, and especially on their observational skills. So far, the use of simplified component state conditions has dominated in maintenance models as the article [8] states. It refers neither to their gradual deterioration and uneven wear nor to unstable ambient conditions.

Spare parts inventories are a problem constant in maintenance management. Their relationship to Industry 4.0 under category (d) will be key. There are problems with the supply of spare parts. They are expensive, lead times are long, and, in some cases, there is a tendency to apply cheaper substitutes. There is also the implementation of fitting the equipment with several sensors for each key component, which hits the cost growth. The maintenance employees in category (c) must change in terms of competencies and qualifications under the influence of Industry 4.0 and Artificial Intelligence, which are more demanding requirements for maintenance employees compared to, for example, the findings in the book [30]. These are different algorithms and patterns of behaviour and thinking that previous generations of maintenance employees are not used to. In addition, maintenance is facing an ageing workforce.

6. Characters in maintenance management

Commonly used in maintenance management are characters (parameters) such as time to failure, repair time, maintenance cost and cost of spare part. Critical characteristics in quality management, as the basis of an integrated management system, including maintenance, are related to the safety of people, equipment, the environment and data. So far, they have generally been underestimated by society. Safety measures for maintenance employees are essential. This is a very risky job. They often work while the machine is running, under voltage or with media in a critical condition (e.g., hot water, hot oil, pressurised steam, volatile substances, etc.).

Critical characteristics of the equipment, as a product, should also be identified in the technical documentation from the designer or technologist. They often correspond to limits of operation of products (objects) and their action in the surrounding environment (e.g., minimum or maximum temperature). They are based on properties of materials used and manufacturing technologies applied. Predictive maintenance should be aware of these characters and focus on them. From these, it should design other maintenance control activities. Production and maintenance planning and control models often contain sets of coefficients that may not correspond to the real system. Reconfiguring them often causes difficulties.

It can destabilise a functioning system or put it outside the legal limits of its functionality (e.g., exhaust emissions). Models usually address ideal objects and obvious scenarios. They reject uncertainties, for example, in the maintenance system, as unrealistic. Of course, it is the unlikely scenarios that start to appear more frequently in turbulent times, but the system is not prepared for them. The use of end-of-life and maintainability components are related. Ignoring these safety limits at the first moment leads to savings in maintenance costs. However, it can usually lead to a safety incident later, which will make the issue considerably more expensive even if no harm to the people in the vicinity (maintainers or other stakeholders) is caused. There are known cases of serious accidents in the history of maintenance where the original cause was the failure to replace or treat a minor component. At present, this phenomenon must be prevented.

In retrospect, these characters and safety/accident rules should by definition be automatically checked when auditing the maintenance management system. Failure to do so is a failure on the part of auditors, as this area is within their remit. Alternatively, auditors should have an expert on that special equipment with them during the audit.

7. Current status of maintenance issues in organisational practice

Observations of maintenance in organisations and analyses of the maintenance management system and the quality management system have revealed many discrepancies. Current maintenance issues in organisations include:

- Incorrectly selected type of maintenance task (confusion between breakdown and preventive maintenance).
- A fault is assigned to a different piece of equipment than the real one (the error may be caused by the fact that it is not clear at the beginning what caused the malfunction).
- The maintenance record cannot be assigned to the device because the device is found to be missing from the register.
- A device is found to be in the register even after it has been removed or replaced by another device.

- Maintenance schedules are not consistent across different information systems or their modules.
- The fault codebook is not used correctly.
- Inconsistency in the number and type of stock items between the information system and the actual situation.
- Selection procedures do not adequately address the quality and reliability of equipment and spare parts.
- The budget of the maintenance section does not consider management's requirement for high availability of equipment functions.

The frequency is not relevant to the above nonconformities. Most of the failures are systemic, not random. It is also not possible to determine the severity because the same problem will have different severity at different facilities and in different organisations. Each device is differently important and therefore the severity of the nonconformities associated with them will be different. Based on the 6 M method, it can be concluded that all these types of nonconformities in maintenance management were originally caused by a human factor. Either they are caused by a maintenance person, a warehouse employee, a maintenance application programmer or some manager. It is also noticeable that the maintenance supply of spare parts is often compromised. Building a functioning information system with up-to-date data is also impaired. All of this can then overlap with the integration of maintenance with Industry 4.0. Above and beyond this, the AI acquires data on the shortcomings of the maintenance system, which then does not always guide its reasoning in an appropriate way. Rather, nonconformities require the attention of employees, their systematic and consistent work with the system and the rules set. Sophisticated maintenance methods are somewhat less relevant in this sense. They may be too robust and costly for operational problem-solving in small and medium-sized organisations. Therefore, it pays them to use proven simpler tools and to pay attention to the diligence of the staff. In addition, the findings of the survey in Ref. [29] should be noted. Preventive maintenance is rather on average to poor condition in organisations in the Czech Republic. Even medium-sized organisations have an unnecessarily organisational and administratively complicated maintenance management system. Positives are the linkage of maintenance to strategy; the care of spare parts and digitalization is developed through different types of sensors connected to the equipment and the associated monitoring of various reliability parameters of the object. On the other hand, the support of extensive administration around maintenance and the form of work orders for maintenance in conjunction with its directive management in a hierarchical organisational structure can be seen as negatives. This in turn leads to an increase in maintenance overhead, long information flows and a decrease in the motivation of maintenance staff. The visualisation side of the maintenance software now dominates over the hardware support (interfaces, interconnects and so on). However, the lack of inclusion of maintenance safety, environmental and information safety and social responsibility aspects in maintenance management systems in organisations is fundamental. The road to predictive maintenance will therefore be a long one.

8. Discussion

Prior historical data on the system are needed for any decision-making and prediction. However, in organisations in the Czech Republic, historical data not only from equipment maintenance is often missing, as in the text [23]. This makes subsequent prediction and planning almost impossible. Also, training AI on this basis is quite difficult.

For example, the CBM method is not used so much in the Czech Republic. However, several other methods mentioned here, such as mathematical-statistical calculations of object reliability, determination of maintenance KPI, the reliability centred maintenance (RCM) method, the multicriteria decision-making, or the HAZOP method are used. This is an indisputable advantage for solving maintenance problems of complex systems that often surround people nowadays.

The fact that a complex system will fail faster due to a large number of components is true from the system theory perspective (as in the book [36]). It includes the individual components as elements, but also their links due to interconnections or other interactions such as vibrations. The individual quality levels of the individual objects build on each other, which contributes to a faster destabilisation of the resulting system.

Binding of the maintenance management model in the study [14] to a hierarchical structure is possible. However, it creates disharmony with respect to organisations that apply the process approach in their management system. Thus, they should have some kind of a heterarchical structure. While the incorporation of various methods into maintenance management is a normal continuation of the development of these system areas and the link to Industry 4.0 is to be expected.

However, linking maintenance management with facility sustainability is a new phenomenon. Part of this is the perceived pressure from the European Union to change the ratio from landfill to end-of-life recycling facilities. On the other hand, there is a deterioration in the degradability of some layered materials into base materials, such as metal-plastic. In the context of facility sustainability, it should be specified that it can be understood as careful maintenance to ensure the long uptime of the facility, or as the production and management of the facility regarding both environmentally sound operation and subsequent recycling.

The determining of causes of nonconformities is not so common in conventional maintenance, outside critical infrastructure. However, it is becoming increasingly important for predictive maintenance purposes. It forms a kind of basis for it. It is only on this basis that the appropriate form of maintenance planning and management can be chosen so that the subsequent predictions are consistent with the reality of the system in its essential environment. This then leads to a more realistic model of the facility maintenance system.

The knowledge and skills of the maintenance staff are becoming a combination of maintenance and computer expert due to the involvement of information technology. This consequently changes the competence model of maintenance employees. In addition, different levels of hierarchy have different scope of authority and responsibility. Furthermore, the skills required are not clear, and there is no clear terminology around Industry 4.0 and AI for a part of the maintenance workforce. In addition, these modern technologies are running up against their limits in real-life organisations, as well as their financial capabilities.

9. Establishing basic trends in maintenance development

The original six trends were further analysed considering findings above and characteristics of the maintenance process. This is based on the quality management system according to the text [32]. Factors are maintenance employee, equipment, spare parts, documentation (the maintenance passport), tools, methods, digitalisation, characters, maintenance costs, legislation and essential environment.

Extending the life of components is a necessity in organisations rather than a need. This is due to the stock disparity of spare parts in many organisations in terms of availability and deteriorated customer-supplier relationships. It should also be noted that extending the life of components should be consistent with their maintainability and sustainability, but at the same time must not compromise the functionality or safety of the equipment and people around. The limit is exceeding the critical lifetime, which leads to the destruction of the system. However, the maintenance cannot be done without spare parts.

The area of maintenance personnel and their competencies should focus on consistency of records, parts drawdown and their timely de-stocking in written documentation and computerised information systems as the production planning and control system (PPC) is. The AI draws from the internet and the company's own information system database. So-called the best maintenance practices are not adequately covered even within the Internet. It will allude to the emotional component of the actions of maintenance personnel. It should help to monitor the consistency of maintenance tasks carried out within the computerised information system or Industry 4.0 and not allow discrepancies to arise. It should also monitor the maintenance person's state of mind and his or her immediate competence (i.e. perception, work speed, error rate, risk-taking and so on).

Therefore, the essential maintenance trends for organisations are modified as:

- Competence of maintenance employee
- Spare parts
- Digitalisation – IOT, PPC, AI and so on.
- Environmental system – but society has not fully come to accept it.
- Methods – often simple rules, procedures and boundaries are sufficient, as well as an obvious definition of maintenance features. Common methods of maintenance management and strategic planning and control are described in detail in books [30, 31], so they are not listed here.

10. Conclusion

As a result of extensive changes in the physical environment, there is an increasing need for equipment maintenance as a form of equipment renewal. At the same time, there are risks that can jeopardise it. Most of these are personnel and systemic. The aim is to extend the life of a whole facility and all its components, usually at low cost. This puts pressure on the planning and management of maintenance and the availability of spare parts. Organisations need to steer their maintenance strategy in the appropriate direction to strengthen their target competitiveness.

Therefore, based on the analyses, key maintenance trends for organisations have been identified. The most key area is the competence of maintenance personnel. This is followed by spare parts and their appropriate management. Digitalisation affects the internal and external context of the organisation. It is the most dynamic factor. It provides several benefits, such as the collection and processing of previously unsuspected amounts of production and maintenance data. However, possible risks lie in its current lack of exploration and imperfection. The environmental system is the answer to the requirement for sustainability in maintenance, as it will cover both degrading equipment and spare parts in terms of their recycling and packaging and the environmental impact of operating substances. The final area of trends is methods suitable for maintenance. Here, it will depend on the requirements and capabilities of the organisation, what they are inclined to do and what they have financial resources to do.

Changes in the essential environment of organisations and their maintenance are very dynamic and therefore trend priorities could shift. However, this list has also been established with a view of its longer-term stability. Furthermore, the planning and management and maintenance methods could not be described in detail within the defined scope.

Acknowledgements

This work was supported with institutional support for long-term strategic development of the Ministry of Education, Youth and Sports of the Czech Republic.

Author details

Věra Pelantová* and Jaroslav Zajíček
Faculty of Mechatronics, Informatics and Interdisciplinary Studies, MTI, OSR, Technical University of Liberec, Liberec, The Czech Republic

*Address all correspondence to: vera.pelantova@tul.cz

References

[1] Morávek D. Velký rozcestník změn v roce 2024, které se týkají podnikatelů. Podnikatel.cz. 2024. Available from: https://www.podnikatel.cz/clanky/velky-rozcestnik-zmen-v-roce-2024-ktere-se-tykaji-podnikatelu/

[2] Dronská J. Záchrana ekonomiky očima krizových manažerů. MM Průmyslové spektrum. 2023;**1**(9):17. ISSN 1212-272

[3] Novák F. Přehled nových technických norem v ČR—leden 2024. Dřevařský magazín. 2024. Available from: https://drevmag.com/cs/2024/01/03/prehled-novych-technickych-norem-v-cr-leden-2024/

[4] Tseremoglou I, Santos BF. Condition-based maintenance scheduling of an aircraft fleet under partial observability: A deep reinforcement learning approach. Reliability Engineering & System Safety. 2024;**241**:109582. ISSN 0951-8320. DOI: 10.1016/j.ress.2023.109582. Available from: https://www.sciencedirect.com/science/article/pii/S0951832023004969

[5] Karabağ O, Bulut Ö, Toy AÖ, Fadıloğlu MM. An efficient procedure for optimal maintenance intervention in partially observable multi-component systems. Reliability Engineering & System Safety. 2024;**244**:109914. ISSN 0951-8320. DOI: 10.1016/j.ress.2023.109914. Available from: https://www.sciencedirect.com/science/article/pii/S0951832023008281

[6] Ghorbani M, Nourelfath M, Gendreau M. Stochastic programming for selective maintenance optimization with uncertainty in the next mission conditions. Reliability Engineering & System Safety. 2024;**241**:109624. ISSN 0951-8320. DOI: 10.1016/j.ress.2023.109624. Available from: https://www.sciencedirect.com/science/article/pii/S0951832023005380

[7] Kamariotis A, Tatsis K, Chatzi E, Goebel K, Straub D. A metric for assessing and optimizing data-driven prognostic algorithms for predictive maintenance. Reliability Engineering & System Safety. 2024;**242**:109723. ISSN 0951-8320. DOI: 10.1016/j.ress.2023.109723. Available from: https://www.sciencedirect.com/science/article/pii/S0951832023006373

[8] Mikhail M, Ouali MS, Yacout S. A data-driven methodology with a nonparametric reliability method for optimal condition-based maintenance strategies. Reliability Engineering & System Safety. 2024;**241**:109668. ISSN 0951-8320. DOI: 10.1016/j.ress.2023.109668. Available from: https://www.sciencedirect.com/science/article/pii/S0951832023005823

[9] Bigi F, Bosi T, Pineda-Jaramillo J, Viti F, D'Ariano A. Long-term fleet management for freight trains: Assessing the impact of wagon maintenance through simulation of shunting policies. Journal of Rail Transport Planning & Management. 2024;**29**:100430. ISSN 2210-9706. DOI: 10.1016/j.jrtpm.2023.100430. Available from: https://www.sciencedirect.com/science/article/pii/S2210970623000628

[10] Gąsieniec L, Jurdziński T, Klasing R, Levcopoulos C, Lingas A, Min J, et al. Perpetual maintenance of machines with different urgency requirements. Journal of Computer and System Sciences. 2024;**139**:103476. ISSN 0022-0000. DOI: 10.1016/j.jcss.2023.103476. Available from: https://www.sciencedirect.com/science/article/pii/S0022000023000818

[11] Sharifi M, Taghipour S. Redundancy allocation problem with a mix of components for a multi-state system and continuous performance level components. Reliability Engineering & System Safety. 2024;**241**:109632. ISSN 0951-8320. DOI: 10.1016/j.ress.2023.109632. Available from: https://www.sciencedirect.com/science/article/pii/S095183202300546X

[12] Hernández MP, Puchkova A, Parlikad AK. Multi-agent learning of asset maintenance plans through localised subnetworks. Engineering Applications of Artificial Intelligence. 2024;**127, Part B**:107362. ISSN 0952-1976. DOI: 10.1016/j.engappai.2023.107362. Available from: https://www.sciencedirect.com/science/article/pii/S0952197623015464

[13] Bhambri P, Rani S. Challenges, opportunities, and the future of industrial engineering with IoT and AI. In: Book Integration of AI-Based Manufacturing and Industrial Engineering Systems with the Internet of Things. 1st ed. London: CRC Press; 2023. p. 18. ISBN 978-10-03-3835-05. Available from: https://www.taylorfrancis.com/chapters/edit/10.1201/9781003383505-1/challenges-opportunities-future-industrial-engineering-iot-ai-pankaj-bhambri-sita-rani

[14] Hlihel BF, Chater Y, Boumane A. Developing a competency model for maintenance 4.0 stakeholders. International Journal of Quality & Reliability Management. 2024, Vol. ahead-of-print No. ahead-of-print. ISSN: 0265-671X. DOI: 10.1108/IJQRM-05-2023-0151 Available from: https://www.emerald.com/insight/content/doi/10.1108/IJQRM-05-2023-0151/full/html

[15] Khazaelpour P, Hashemkhani ZS. FUCOM-optimization based predictive maintenance strategy using expert elicitation and artificial neural network. Expert Systems with Applications. 2024;**238, Part A**:121322. ISSN 0957-4174. DOI: 10.1016/j.eswa.2023.121322. Available from: https://www.sciencedirect.com/science/article/pii/S0957417423018249

[16] Adam R. Průmysl 4.0 ve společnosti Windmöller and Hölscher. Automa. 2023;**29**(6):39-40. ISSN 1210-9592

[17] Kulshrestha N, Agrawal S, Shree D. Spare parts management in industry 4.0 era: A literature review. Journal of Quality in Maintenance Engineering. 2024. DOI: 10.1108/JQME-04-2023-0037 Vol. ahead-of-print, No. ahead-of-print. Available from: https://www.emerald.com/insight/content/doi/10.1108/JQME-04-2023-0037/full/html

[18] Samadhiya A, Agrawal R, Garza-Reyes JA. Integrace průmyslu 4.0 a celková produktivní údržba pro globální udržitelnost. The TQM Journal. 2024;**36**(1):24-50. DOI: 10.1108/TQM-05-2022-0164. Available from: https://www.emerald.com/insight/content/doi/10.1108/TQM-05-2022-0164/full/html

[19] Kassay Š. Promluvy Štefana Kassaye. Sváteční gratulace ve znamení umělé inteligence. MM Průmyslové spektrum. 2023;**1**(12):37. ISSN 1212-272

[20] Janišová H, Vaněk J. Šestkrát chytrý facility management. MM Průmyslové spektrum. 2023;**1**(10):64-65. ISSN 1212-272

[21] Beckmann N. Inteligentní pneumatika: brána k vyšší účinnosti a produktivitě. Automatica. 2023;**29**(5):44-46. ISSN 1210-9592

[22] Abbas AN, Chasparis GC, Kelleher JD. Hierarchical framework

for interpretable and specialized deep reinforcement learning-based predictive maintenance. Data & Knowledge Engineering. 2024;**149**:102240. ISSN 0169-023X. DOI: 10.1016/j.datak.2023.102240. Available from: https://www.sciencedirect.com/science/article/pii/S0169023X23001003

[23] Milena N, Fruggiero F, Lambiase A, Bruton K. A systematic mapping of the advancing use of machine learning techniques for predictive maintenance in the manufacturing sector. Applied Sciences. 2021;**11**(6):2546. DOI: 10.3390/app11062546 Available from: https://www.mdpi.com/2076-3417/11/6/2546

[24] Dhanola A, Garg HC. Tribological challenges and advancements in wind turbine bearings: A review. Engineering Failure Analysis. 2020;**118**:104885. ISSN 1350-6307. DOI: 10.1016/j.engfailanal.2020.104885. Available from: https://www.sciencedirect.com/science/article/pii/S1350630720314096

[25] Jambadu L, Monstadt J, Pilo F. The politics of tied aid: Technology transfer and the maintenance and repair of water infrastructure. World Development. 2024;**175**:106476. ISSN 0305-750X. DOI: 10.1016/j.worlddev.2023.106476. Available from: https://www.sciencedirect.com/science/article/pii/S0305750X23002942

[26] Kamel E, Habibi S, Memari AM. State of the practice review of moisture management in residential buildings through sensors. Structures. 2024;**59**:105698. ISSN 2352-0124. DOI: 10.1016/j.istruc.2023.105698 Available from: https://www.sciencedirect.com/science/article/pii/S2352012423017861

[27] Lilli G, Sanavia M, Oboe R, Vianello C, Manzolaro M, De Ruvo RL, et al. A semi-quantitative risk assessment of remote handling operations on the SPES front-end based on HAZOP-LOPA. Reliability Engineering & System Safety. 2024;**241**:109609. ISSN 0951-8320. DOI: 10.1016/j.ress.2023.109609. Available from: https://www.sciencedirect.com/science/article/pii/S0951832023005239

[28] Hutchinson B, Dekker S, Rae A. Audit masquerade: How audits provide comfort rather than treatment for serious safety problems. Safety Science. 2024;**169**:106348. ISSN 0925-7535. DOI: 10.1016/j.ssci.2023.106348. Available from: https://www.sciencedirect.com/science/article/pii/S0925753523002904

[29] Hroch J, Šimek V, Pexa M. Prediktivní údržba—zkušenosti z pohledu auditů údržby. Vše o průmyslu.cz. 2022. Available from: https://www.vseoprumyslu.cz/udrzba-a-diagnostika/asset-management/prediktivni-udrzba-zkusenosti-z-pohledu-auditu-udrzby.html

[30] Legát V et al. Management a inženýrství údržby. 2nd ed. Příbram: Karel Mařík—Professional Publishing; 2016. ISBN 978-80-7431-163-5

[31] Madzík P. Nástroje systematického riešenia problémov. 1st ed. Ružomberok: VERBUM; 2017. ISBN 978-80-561-0478-1

[32] Standard ISO 9001. Quality Management Systems—Requirements. Prague: UNMZ; 2016

[33] Standard ISO 14001. Environmental Management Systems—Requirements and Instructions for Use. Prague: UNMZ; 2016

[34] Standard EN 13306 Maintenance—Maintenance Terminology. Prague: UNMZ; 2018

[35] Standard ISO/IEC 27000. Information Technology—Security New Maintenance Management Topics. Techniques—Information Security Management Systems—Overview and Specialist Dictionary. Prague: UNMZ; 2020. DOI: 10.5772/intechopen.100515517

[36] Pelantová V, Havlíček J. Integrace a systémy Management. Monografie. Edition 1st ed. Liberec: TUL, FM; 2014. ISBN 978-80-7494-164-1

[37] Mykiska A, Legát V, et al. Zabezpečování spolehlivosti. Prague: ČSpÚ; 2001

Chapter 3

Maintenance Execution: What and How – A PDCA Approach

Christian Okonta, Ralphael Edokpia and Christopher Eboigbe

Abstract

Maintenance execution determines the overall outcome of any maintenance policy and strategy adopted to keep a facility in a reliable state. Once this is flawed, every other aspect of the maintenance process will not yield the intended result, hence the need to focus attention on the shopfloor execution of maintenance tasks with adherence to plans and schedules. There are two major aspects of maintenance execution: what to do and how to do it. What to do are a combination of original equipment manufacturer recommendations and other source of information relevant to maintenance planning, such as historical data, experience from technicians, and prognostic methodologies for data gathering. While this is almost unique to equipment, it is a prerequisite for proper maintenance planning and execution. The how-to-do aspect of maintenance execution involves the capability and ability of the shopfloor technicians to carry out the maintenance plan with the right knowledge, attitude, and tool set. By adopting the PDCA approach, a systematic approach to maintenance execution is developed that covers both planning and implementation of maintenance execution for sustaining reliability. The result shows a steady decline in waste trend with over 5% reduction in the amount of waste in less than 3 months.

Keywords: maintenance, maintenance execution, PDCA, basic condition, preventive maintenance

1. Introduction

The aim of maintenance is to keep equipment function in basic condition defined by safety, quality, and productivity constraints. Continual usage of equipment leads to deterioration which can propagate to defect and subsequent breakdown. To detect and eradicate defects and anomalies, maintenance is carried out. Thus, maintenance is catholic term used to describe all activities carried out on a piece of equipment to prevent it from breaking down and resulting in unplanned downtime [1]. It could be cleaning, inspection, lubrication, tightening, adjustment, repairs or replacement.

In the manufacturing sector, maintenance cost is usually a major concern to the stakeholders, and optimization of this cost is a key performance index of the maintenance management team [2]. The ability to predict critical failures that may lead to long downtime is a key factor in reducing the overall cost along the supply chain [3]. Adequate maintenance is necessary as any malfunction that arises during manufacturing would cause a disturbance in the supply chain [4].

Different organizations adopt different policies for maintenance depending on the criticality of the expected failure. Preventive maintenance policy is usually implemented where equipment failure could result in huge downtime and production loss. In this situation, the cost of failure is high, and the aim is to reduce unplanned downtime to the minimum. Corrective maintenance policy is mostly common when the effect of failure is minimal, and the repair time is usually short or negligible; hence, no serious investment is made on inspection. In some production companies, there is a combination of both preventive and corrective maintenance policies which are implored to keep different machines in running state based on equipment complexity and criticality. The method of defect identification and determination of remaining useful life of an equipment is referred maintenance techniques, while the application maintenance policy and techniques to keep the facility functional is called maintenance strategy [5]. According to a study by Wang et al., [6] three competing failures may occur in a system, namely: a) shock failure as a result of environment shocks, b) soft failure which occurs when the deterioration is allowed to exceed a critical value, and c) out of balance when the difference value in alignment among components reaches the failure threshold. The ability to predict critical failures that may lead to long downtime is a key factor in reducing the overall cost along the supply chain [7]. Any breakdown that occurs during production will lead to a disruption in the supply chain, hence the need for adequate maintenance [8].

To create a good preventive maintenance system, a prerequisite is a standard roadmap to detail various action plans for maintenance execution with emphasis on what needs to be maintained and how the maintenance should be carried out. This is the process of finding the best possible maintenance strategy for every asset in your organization with the end goal of achieving consistently high levels of reliability at the lowest possible costs.

This work is structured in sections. Section 2 explains the WHAT of maintenance execution, while Section 3 gives a brief of the HOW. Section 4 elaborates on the adopted PDCA road map used to address the question arising from the WHAT section and also explains the HOW approach. Section 5 is the result obtained from the implementation of the approach using a food production factory as a case study. The results obtain are discussed in Section 6, while Section 7 is the conclusion.

2. Maintenance execution: what

Based on the principle of reliability-centered maintenance, understanding what needs to be done in order to keep equipment in a sustainable reliability state sets the maintenance process of identification of failure modes in motion for successful eradication through countermeasures. To get started, these questions must be addressed.

1. What is the working principle of the asset or equipment, and what are the associated performance standards?

2. What are the assemblies, components and parts of the machine that require maintenance? They are referred to as maintenance significant items (MSI).

3. In what ways can the maintenance significant items fail to provide the required functions? This involves failure mode identification.

DOI: http://dx.doi.org/10.5772/intechopen.1005270

4. What event can trigger each failure?

5. What are the noticeable signs and possible effects when each failure occurs?

6. What are the risks of failure?

7. What systematic proactive task can be done to prevent or diminish the consequence of the failure?

3. Maintenance execution: how

The next part in the maintenance execution journey is "How" maintenance should be carried out on the recognized items in order to sustain the reliability of the equipment at a standard safety and quality level. This is an explanation of required procedures in carrying out the WHAT of the maintenance execution. The two major aspects of the How in maintenance execution are the type and nature of maintenance required by these items. The type of maintenance could be time-based, condition-based, or prescriptive-based preventive maintenance depending on the accessibility, difficulty, and criticality of the machine/component. The nature of the maintenance is the actual action that needs to be performed on the maintenance significant item for accurate defect detection and restoration. This is where the failure modes of the component are being looked out for. The best approach is by asking questions around the possible effect of the identified failure modes on the MSI. The response to these questions together with recommendation from original equipment manufacturer (OEM) and historical experience becomes an insight into the creation of task list for inspection and maintenance. The next part is to prepare a detailed instruction and procedure for the WHAT part earlier identified with required tools, manpower, and enough training guide for technicians. For validation purpose, trigger the preventive maintenance as a corrective maintenance action and watch the execution. This second step is an opportunity to test your process and make possible adjustments before rolling out your maintenance task lists as a standard inspection and maintenance plan.

4. Maintenance execution PDCA roadmap

Maintenance is a continuous process and thus requires continuous improvement. In the 1950s, W. Edwards Deming developed a quality system for the continuous improvement of business processes. While Deming concentrated on industrial production processes, his approach and ideas could be applied just as readily for maintenance and other contemporary corporate operations [9]. Deming developed a model for use for constant improvement known as the PDCA (Plan-Do-Check-Act) cycle [10].

Implementation of the PDCA cycle for maintenance system execution is particularly important for the maintenance manager to remain focus on the basic of the maintenance program and not carried away by issues on the shop floor. It systematically provide answers to the relevant questions asked in the 'WHAT' section of maintenance execution, addressing each of them as a standard work process to developed and cascaded to the shopfloor. It also elaborates on the 'HOW' process of proactive maintenance execution. The implementation of the PDCA cycle for sustainable maintenance execution is elaborated in the following below.

4.1 Plan

Defining the issue or topic is the first step in finding long-term fixes. Creating an action plan, including the necessary resources and planning, is the final step in this process. This could be referred to as standard work process development of the maintenance system.

The first step is to develop preventive and corrective maintenance work order management route as shown in **Figures 1** and 2.

The maintenance work order management route in **Figures 1** and **2** is a definition of a systematic approach to upkeep the facilities and equipment, and it varies from facility to facility [11]. The selection of the best approach to sustaining a particular asset has been quite a challenge for the asset owner. The choice of the best maintenance strategy is a selection of the most suitable maintenance techniques for the policy to be operationalized. Maintenance techniques (MT) in this context are the methods used to forecast and predict the remaining useful life of an asset [12]. MT enables the application of maintenance policies. The available data as well as the possibilities for data collection and the required outcome determine what MT to select. Based on the classification [10, 11], five different forms of MTs can be distinguished as follows.

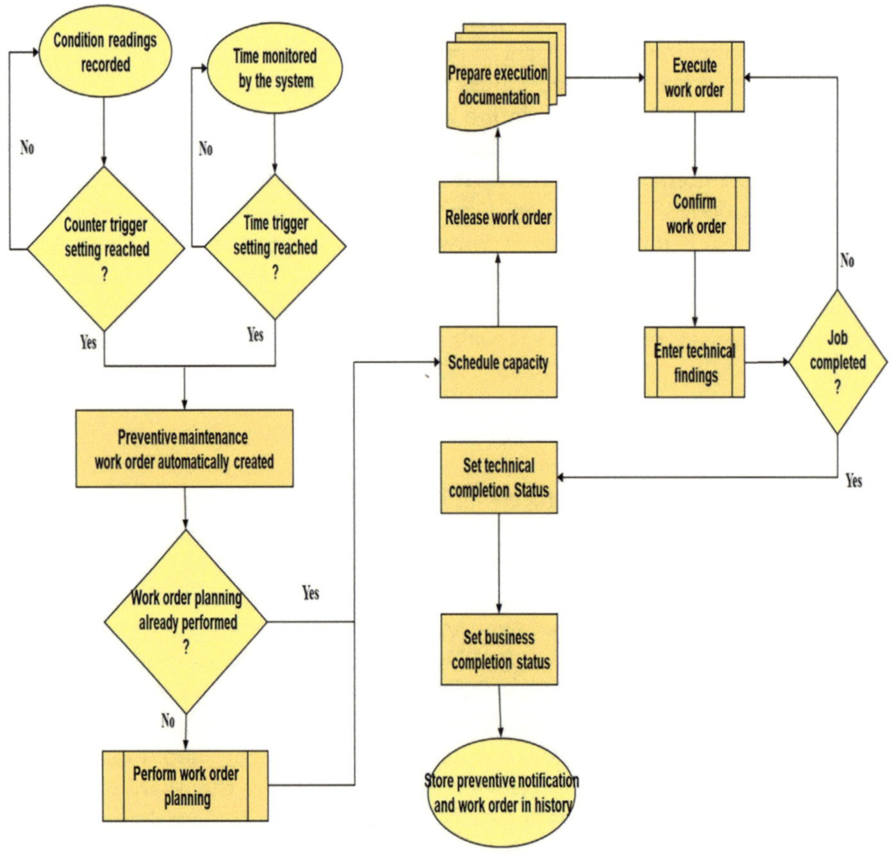

Figure 1.
Preventive maintenance work order management route.

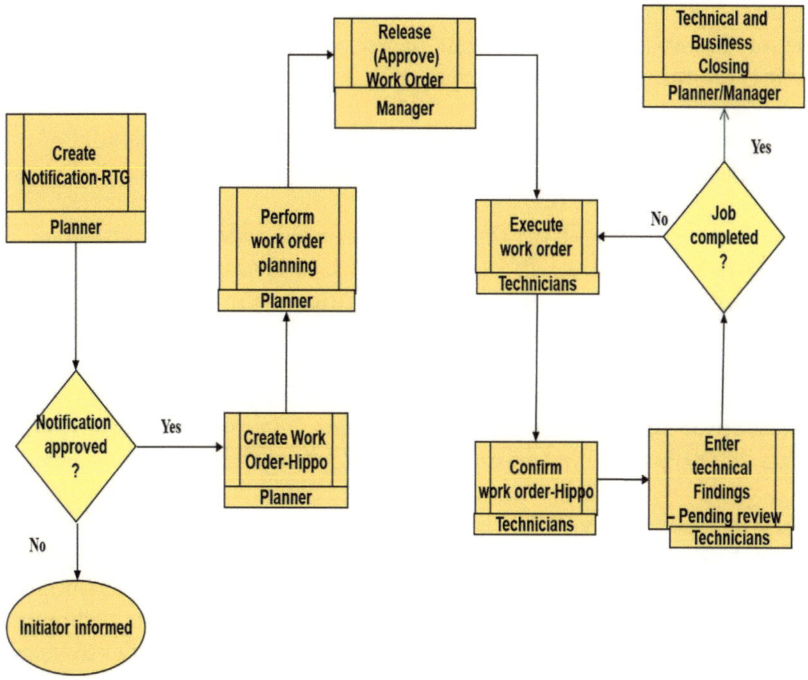

Figure 2.
Corrective maintenance work order management route.

i. Experience-based: in the experience-based technique, forecasts of failure times are based on knowledge and experience outside or inside the organization (e.g., OEM). Occasionally, little or sparse data support them. Predictions are based on expert judgment (for example, facilitated by FMECA techniques). Such methods estimate the life span of an average variable operating under average historical conditions [13–15].

ii. Reliability statistics: The prediction techniques for reliability statistics are based on historical (failure) records of comparable equipment without regard for specific component (use) variation. This method explains precisely the likelihood of population-wide failure. Such techniques also estimate the lifespan of an average item that operates under average conditions, for example, Distributions of Weibull [16].

iii. Stressor based: stressor-based predictions are based on historical records with stressor data, for example, temperature, moisture, or speed, including environmental and operational variances, and produce predicted system life expectancy results in a specific environment. Pronouncements are based on a general direction derived from a physical model, built-in performance, or operating history [13].

iv. Degradation based: degradation-based prognosis is based on the extrapolation of a general path to a failure threshold from a forecast parameter, a degradation measure. The system can be diagnosed by measuring symptoms of initial failure such as an increase in temperature or vibration. The prognostic parameter is also calculated by sensor measurements that are often dependent on measurement. The forecast starts from the current deterioration situation and results in a lifetime required of a particular system in a certain environment [17].

v. Model-based: model-based projections give the predicted remainder of the lifetime of a certain system. Two distinct categories of models exist:

a. Physical model-based: the prognostic parameter is computed based on a degradation mechanism physical model based on direct load-sensing or usage which governs individual components' critical failure mechanisms.

b. Data model-based: data analytics approach that uses sensed load variations, utilization of process data or condition/health monitoring data as input to measure or extract the prognostic parameter. The algorithms are designed to extract or try to predict anomalies by comparing them with historical data.

The aim is to generate value adding preventive and corrective maintenance work orders to keep the equipment in basic condition. Automatic triggers such as condition monitoring and time base counter follows the preventive maintenance workorder management route in **Figure 2**, while work request from technicians and operators' inspections follows the corrective maintenance work order management route in **Figure 3**.

Even with the best maintenance strategy, breakdown is almost inevitable but could be minimized through proper countermeasure. Thus, the next step in the planning phase of the PDCA cycle is to develop breakdown analysis route as shown in **Figure 3** to ensure that breakdowns are not recurrent. The seriousness level of stoppage is determined by the company standard, and any downtime that is equal to or greater than the defined standard (say 30 minutes) is classified as a serious breakdown and thus selected for analysis. It is recommended to employ the Five Why (5 whys) problem-solving technique, which investigates the underlying causes and effects of specific issues. The main objective is to ask "Why?" repeatedly in order to identify the underlying source of a flaw or issue. Failures will recur if incorrect conclusions are drawn from breakdown data analyses that do not take into account the underlying processes of failure. Before any breakdown may be permanently eliminated by the application of suitable countermeasures, the failure mode of that breakdown must be connected to the root cause.

4.2 Do

This is the daily implementation of the planned actions. The objective of this step is to implement the plan defined above. The first action is the resource allocation.

Resource allocation is the process of assigning and managing assets in a manner that supports an organization's strategic planning goals while also conferring a fitness benefit by contributing to a defined objective [18]. Resource allocation includes managing tangible assets such as hardware to make the best use of softer assets such as human capital. Resource allocation involves balancing competing needs and priorities and determining the best course of action to maximize the use of limited resources and get the best return on investment [19].

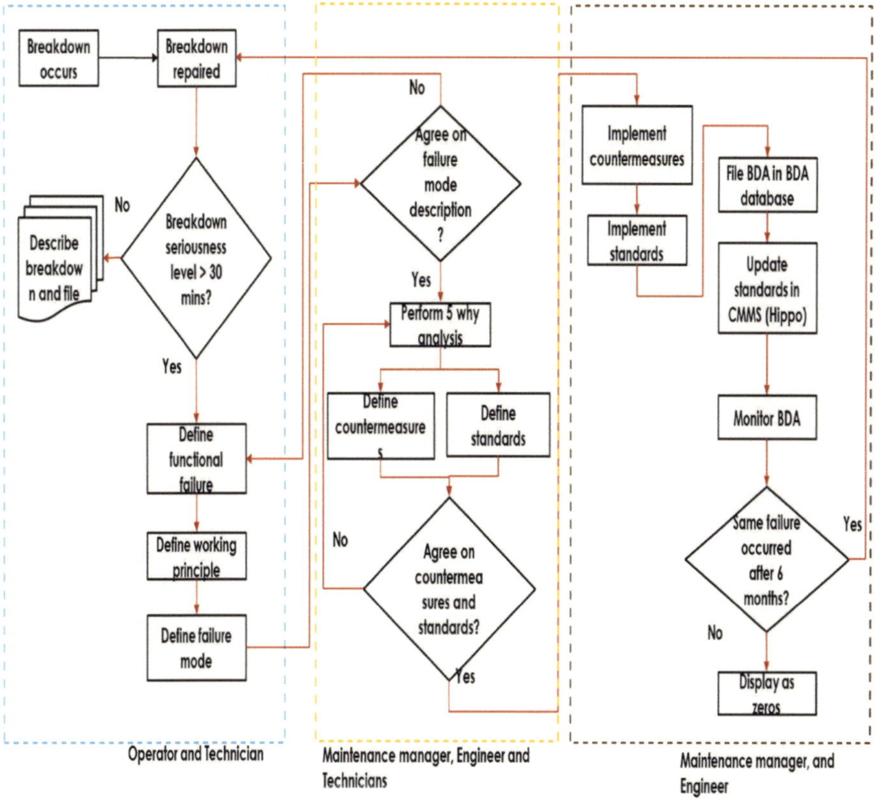

Figure 3.
Breakdown analysis route.

For illustration, using a premium tortilla production factor as a case study, the operational sequence is as follows;

Flour is received from the trucks delivered from the manufacturer to the plant into different silos with loadcells to determine the quantity and level of flour in each silo. Different flour such as whole wheat or white flour are stored in different silos. Once the facility is ready for production, the desired recipe is entered into the human-machine interface of the hopper. The hopper initiates ingredient call from the silo through the vacuum pump. Once the preset quantity is achieved, the vacuum pump cuts off. The ingredient is emptied from the hopper into the mixing bowl which is in turn clamped onto the mixing machine duck for mixing in order to achieve a homogenous quality dough at right elasticity, texture, and temperature. The prepared dough from the mixer is moved to the bowl lift which helps in conveying the bowl on the divider. The divider helps cut the dough into small oval shape of desired size and weight according the installed pocket of the divider drum. The next machine on the line is the proofer. The function of the proofer is to provide a controlled environment with respect to temperature and humidity for exothermic reaction of the enzymes on the dough and provide the right toughness for easy pressing. The autoloader helps to arrange the dough in arrays on the conveyor in preparation for pressing. The press is used to spread the dough ball into a circular flat tortilla of consistent size, ready for baking in the oven. The oven is a temperature-controlled baking system used to cook and bake the tortillas at desired temperature and time. The cooler is a long conveyor in a temperature-controlled room to reduce the temperature of the tortilla coming from the oven to less than 5 degree Celsius. The inspection/rejection system is a quality control installation that helps to inspect the shape, size, consistency, presence of spots and holes, and subsequent reject bad tortillas based on the specified pixel parameter. Counter stacker helps to count the good tortillas passed from the inspection system and stack them into different pockets according to the defined number ready for packaging. The indexing machine is used to press down the stacked tortillas. By pressing it together, it is easier to be transported on the conveyor and much easier for the bagging machine to handle. The function of the bagger is to put the pre-arranged stacks of tortillas into bags. Next is the metal detector, a critical control point for detecting the presence of metal in the tortilla by measurement of the electromagnetic radiation from the stacks. The printer is used for date and batch coding of the bags before being put into cartons and sent for delivering to the final consumers.

A summary of the flow chart of the production process with assigned mechanics is as shown in **Figure 4**.

The next step is the implementation of preventive maintenance daily agenda defining what must be done before the close of each business day. A guide to the maintenance team as shown in **Figure 5**.

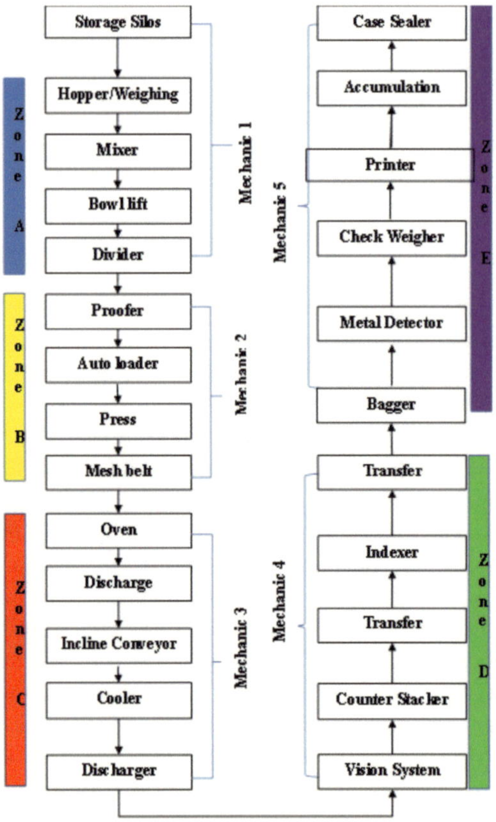

Figure 4.
Resource allocation in a production plant.

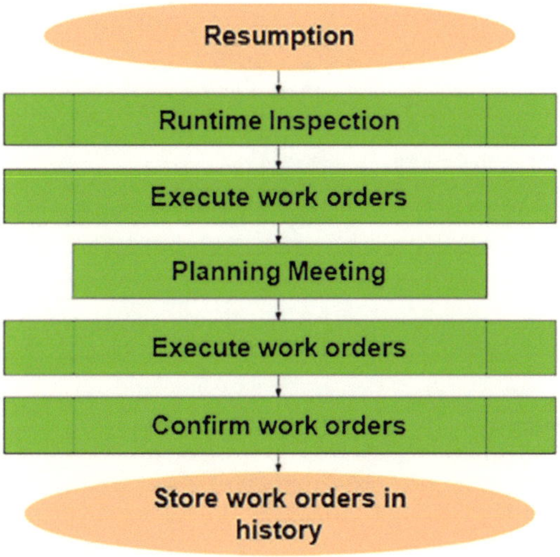

Figure 5.
Preventive maintenance daily agenda.

This agenda serves as a workflow for managing the maintenance team on the shopfloor. It specifies the actions that must be performed on a daily basis to keep the system functional, responsive, and organized. **Figure 6** states that runtime inspection must be carried out by technicians upon resumption followed by the execution of available workorders. These two actions serve as a strong preparation for the daily maintenance planning meeting which present an opportunity to get update from the previous 24 hours, follow-up with planned actions for the next 24 hours, and get update and feedback from the team as in **Figure 6**. Other maintenance scope meetings such as weekly, bi-weekly, and project planning meetings could also be integrated for forward planning. Work order execution is the aim of the "do" phase of the PDCA and must be given more attention. Time must be allotted for the documentation of executed workorders.

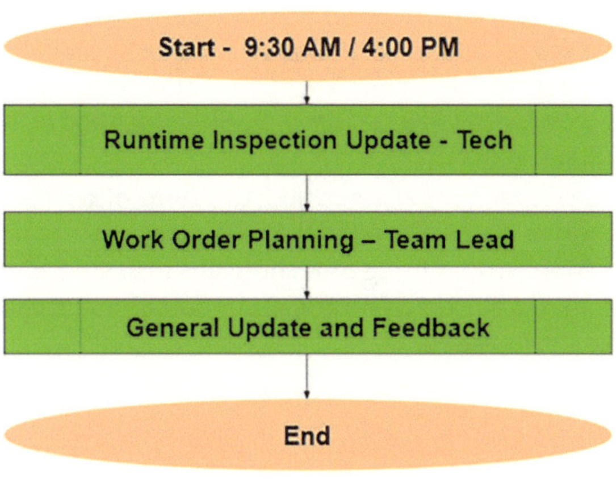

Figure 6.
Daily meeting critical agenda.

Downtime <1000 minutes	
Percentage of Preventive maintenance of Corrective maintenance: PM 70%, CM 30%	
Conformance to PM schedule: 80%	
Work order age	PM <10 days On demand <10 days
Backlog	PM <15 On demand<15

Table 1.
Standardize maintenance metrics.

4.3 Check

CHECK Assess the measurements and report the results to the decision makers. Using measures, indicators, or observations, the effectiveness of measures implemented must be verified. If any modifications are needed, we return to the planning step. Maintenance metric should be clearly stated with target and explained during the weekly KPI review meeting. This objective metrics helps in visualizing the team performance on daily, weekly, or monthly review. **Table 1** shows a standard metric for maintenance performance tracking.

4.4 Act

This is the finalization of results, and their sustainable implementation involves development or updating of documents such as procedures, processes, good practice guides, or forms. ACT decides on the changes that are needed to improve the process. In maintenance, this basically involves review and update of task lists and training of technicians on inspections and maintenance. As a maintenance manager, requirements from maintenance technicians must be clearly stated as a written and documented work instructions. Maintenance work instruction can thus be defined as a written set of instruction that specifies how a maintenance task is to be performed and expected quality from the work order. The reason for work instruction is to manage/limit human error, reduce variability in task performance and ensure adherence to safety procedures. Maintenance work instruction is a living document that is subject to continuous improvement in a bit to finding the most efficient and safest way to perform the given task.

Focusing on the Dixon mixer for processing dry flour into consistent dough as a case study, the preventive maintenance of the mixer is written in detail in **Table 2**.

This can be used for onsite training of technicians to ensure adept understanding of the working principle of the machine and execution of maintenance function as stipulated by the work instruction.

5. Result

Using the case study of a premium food production plant in Canada as referenced in the DO section of the PDCA, the aim is to reduce waste from an average of 25% to

Procedure for monthly mixer maintenance

MSI	Tasks	Picture
1. Agitator assembly	a. Lift the mixer head and install the safety bar 1. Re-torque bolts on spider hubs. 2. Check for any sign of agitator-to-bowl interference. 3. Check agitator for side-to-side motion. 4. Check agitator bar bushing for wear. 5. Check roller bar mounting bolts for wear. 6. Replace when agitator bar bushings are replaced. 	
2. Bowl mounting	Instruction: a. Lift the mixer head and install the safety bar 1. Re-torque all bolts. 2. Inspect all seals for damage or wear. Replace as needed. 3. Grease Trunnion Rings (Tilt Bowl mixers only).	
3. Canopy	Instruction: a. Lift the mixer head and install the safety bar b. Perform the following task: 1. Check flour gate system for excessive wear. 2. Inspect all bowl seals for wear, cracks, cuts, and tears. Replace as needed. 3. Check for airline leaks.	

Procedure for monthly mixer maintenance

MSI	Tasks	Picture
4. Gear assembly	Instruction: a. Lower the mixer head to horizontal position b. Open the top cover of the mixer c. Perform the following task: 1. Check oil level and quality. Change contaminated, burned, or waxed oil. 2. Grease the bearing 3. Check the condition and tension of the belts 	 
5. Hydraulic system	Instruction: a. Lift the mixer head and install the safety bar b. Open the back cover of the mixer c. Perform the following task: 1. Inspect and grease rod ends of the Hydraulic Tilt Cylinder. 2. Check the Hydraulic Tilt System for smooth raising and lowering, pressure settings. 3. Check the oil level is toward the upper limit of the dip stick 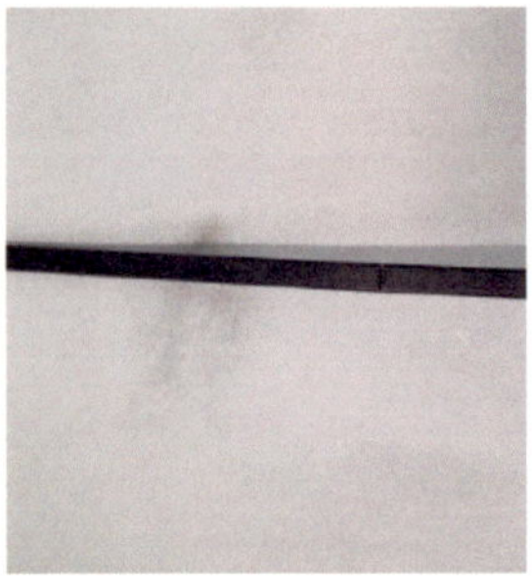	 

Procedure for monthly mixer maintenance		
MSI	**Tasks**	**Picture**
6. Field devices	a. Lift the mixer head, and install the safety bar b. Remove the back cover c. Perform the follow task: 1. Inspect the condition of the four limit switches for actuation and damages 	 

Table 2.
Procedure for monthly mixer maintenance.

less than 10% by implementation of a world class preventive maintenance strategy. By following this maintenance execution approach in three months, the result of the waste trend is shown in **Table 3** and **Figure 7**.

The result from the check on the lagging and leading maintenance responsible for this reduction in waste is shown in **Tables 4** and **5**.

Week of	Waste	Four-week avg.
5-Sep	26.00%	
12-Sep	18.40%	
19-Sep	19.80%	
26-Sep	23.90%	22.0%
3-Oct	20.80%	20.7%
10-Oct	19.20%	20.9%
17-Oct	13.60%	19.4%
24-Oct	15.80%	17.4%
31-Oct	18.80%	16.9%
7-Nov	16%	16.1%
14-Nov	16.40%	16.8%
21-Nov	12.70%	16.0%
28-Nov	12.53%	14.4%

Table 3.
Waste trend data.

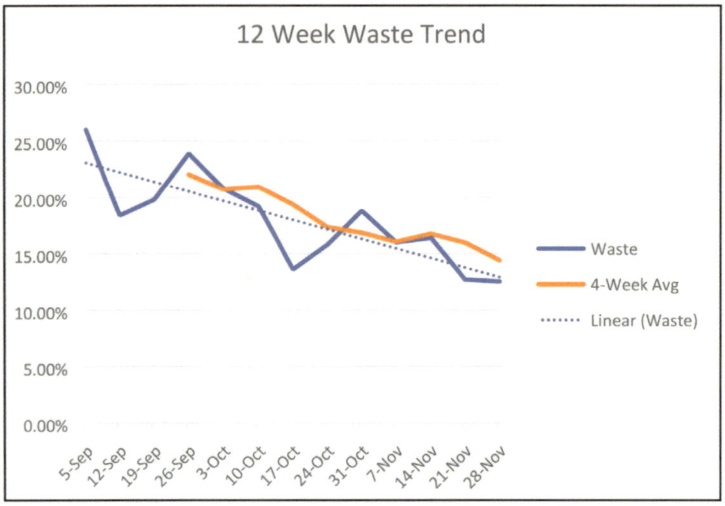

Figure 7.
Three-month waste trend.

Downtime	Actual (minutes)	Target (minutes)
	982	<1000

Table 4.
Maintenance lagging metric.

Percentage of PM to CM (last week)			
Type of maintenance	**Completed**	**Percentage**	**Target**
Preventive maintenance	32	52%	>70%
Corrective maintenance	30	48%	<30%
Conformance to PM (YTD)	**Percentage completed on time**		**Target**
PM	85%		>90%
CM	91%		>95%
Pending YTD		**Target**	
PM	89	<20	
CM	56	<15	
Backlog until 18th-Mar-2023		**Target**	
PM	18	<15	
CM	16	<15	
Work order age		**Target**	
All work orders	7 avg. days	<5 days	

Table 5.
Maintenance leading metric.

DOI: http://dx.doi.org/10.5772/intechopen.1005270

6. Discussion

By following this maintenance execution approach, after 3 months, all the registered equipment of the computerized maintenance management system had been updated with weekly, bi-weekly, monthly, quarterly, semiannual, and annual maintenance plan. The maintenance metrics developed showed an improvement in waste from 25% to less than 15% within 3 months of introduction and adoption of the approach. The maintenance backlog was also reduced to less than 40. By adopting the PDCA approach, a systematic approach to maintenance execution is developed that covers both planning and implementation of maintenance execution for sustaining reliability. The CHECK shows the immediate performance of the maintenance function by elaborating on both the lagging and leading metrics in **Tables 3** and **4**. The result shows a steady decline in waste trend with over 5% reduction in the amount of waste in less than 3 months.

7. Conclusion

This chapter elaborates the PDCA approach to maintenance execution, focusing on what needs to be done and how maintenance execution should be carried out for the identified maintenance significant items. In the plan phase of the cycle, the preventive maintenance road map was presented for effective action plans based on the adoption of different maintenance techniques. The maintenance technique was used as a means of forecasting the remaining useful life of equipment hence, a trigger for the preventive maintenance workorder. A corrective maintenance route was also introduced to handle corrective work order request from inspections. As breakdown is almost inevitable, a breakdown route based on root cause analysis was introduced. The implementation phase focused on defined workflow is to be followed, specifying the maintenance team daily agenda. Standard leading and lagging maintenance metrics were presented as checks, guiding the maintenance function performance. As maintenance is a continuous process, work instruction must be revised and improved upon for onboarding and training technicians. The result of the implementation of this approach to maintenance is the establishment of a self-governing self-sustaining preventive maintenance system that is proactive in defect identification and eradication. The PDCA loop is able to carry out self-assessment of the system maintenance function performance, develop action plan for improvement, and evaluate the impact of the actions on the overall system.

Author details

Christian Okonta*, Ralphael Edokpia and Christopher Eboigbe
University of Benin, Benin City, Nigeria

*Address all correspondence to: okonta.christian@yahoo.com

References

[1] Christer AH, Wang W, Sharp J, Baker R. A case study of modelling preventive maintenance of a production plant using subjective data. Journal of the Operational Research Society. 1998;**49**(3):3. DOI: 10.1057/palgrave.jors.2600518

[2] Zhao X, Nakagawa T. Optimal periodic and random inspections with first, last and overtime policies. International Journal of Systems Science. 2015;**46**(9):9. DOI: 10.1080/00207721.2013.827263

[3] Lin Y-K, Chang P-C, Chen JC. Performance evaluation for a footwear manufacturing system with multiple production lines and different station failure rates. International Journal of Production Research. 2013;**51**(5):5. DOI: 10.1080/00207543.2012.718451

[4] Zhao X, Al-Khalifa KN, Magid Hamouda A, Nakagawa T. Age replacement models: A summary with new perspectives and methods. Reliability Engineering & System Safety. 2017;**161**:95-105. DOI: 10.1016/j.ress.2017.01.011

[5] Ascher H, Feingold H. Repairable systems reliability: Modeling, inference, misconceptions and their causes. In: Lecture Notes in Statistics, Vol. 7. New York: M. Dekker; 1984

[6] Wang J, Liu H, Lin T. Optimal rearrangement and preventive maintenance policies for heterogeneous balanced systems with three failure modes. Reliability Engineering & System Safety. 2023;**238**:109429. DOI: 10.1016/j.ress.2023.109429

[7] Lin Y-K, Chang P-C. A novel reliability evaluation technique for stochastic-flow manufacturing networks with multiple production lines. IEEE Transactions on Reliability. 2013;**62**(1):1. DOI: 10.1109/TR.2012.2220898

[8] Okonta C, Edokpia R. Evaluation of elements affecting the effectiveness of maintenance policies following the ASME code and delay time model. In: Proceedings of the 3rd African International Conference on Industrial Engineering and Operations Management, Nsukka, Nigeria. 5 April 2022. DOI: 10.46254/AF03.20220349

[9] Nie X, Lin M, Xu S, Zhang L, Lin X, Huang W. Strategically reducing carbapenem-resistant Acinetobacter baumannii through PDCA cycle-driven antibiotic management. Indian Journal of Medical Microbiology. 2024;**48**:100527. DOI: 10.1016/j.ijmmb.2024.100527

[10] Von Rosing M, Scheer A-W, Zachman JA, Jones DT, Womack JP, Von Scheel H. Phase 3: Process concept evolution. In: The Complete Business Process Handbook. Vol. I. Amsterdam: Elsevier Science; 2015. pp. 37-77. DOI: 10.1016/B978-0-12-799959-3.00003-3

[11] Mancuso A, Compare M, Salo A, Zio E. Optimal prognostics and health management-driven inspection and maintenance strategies for industrial systems. Reliability Engineering & System Safety. 2021;**210**:107536. DOI: 10.1016/j.ress.2021.107536

[12] Campbell JD, Reyes-Picknell JV. UPTIME Strategies for Excellence in Maintenance Management. 2nd ed. New York: Productivity Press. References - Scientific Research Publishing. [Online]. Available from: https://www.scirp.org/%28S%28vtj3fa45qm1ean45vvffcz55%29%29/

reference/referencespapers.aspx?referenceid=1154996 [Accessed: June 01, 2021]

[13] de Jonge B, Teunter R, Tinga T. The influence of practical factors on the benefits of condition-based maintenance over time-based maintenance. Reliability Engineering & System Safety. 2017;**158**:21-30. DOI: 10.1016/j.ress.2016.10.002

[14] Coble J, Hines J. Applying the general path model to estimation of remaining useful life. International Journal of Prognostics and Health Management. 2011;**2**:2153-2648

[15] Sayed-Mouchaweh M. Artificial Intelligence Techniques for a Scalable Energy Transition: Advanced Methods, Digital Technologies, Decision Support Tools, and Applications. Switzerland: Springer Nature; 2020

[16] Filz M-A, Langner JEB, Herrmann C, Thiede S. Data-driven failure mode and effect analysis (FMEA) to enhance maintenance planning. Computers in Industry. 2021;**129**:103451. DOI: 10.1016/j.compind.2021.103451

[17] Tiddens WW. Setting Sail towards Predictive Maintenance: Developing Tools to Conquer Difficulties in the Implementation of Maintenance Analytics (PhD). Enschede, The Netherlands: University of Twente; 2018. DOI: 10.3990/1.9789036546034

[18] Baghdassarian HM, Lewis NE. Resource allocation in mammalian systems. Biotechnology Advances. 2024;**71**:108305. DOI: 10.1016/j.biotechadv.2023.108305

[19] Ben L. Resource Allocation [Online]. Available from: https://www.techtarget.com/searchcio/definition/resource-allocation

Chapter 4

Optimizing Spare Part Management for Vessels in Liner Shipping

Arameh Bisadi, Amir Zare and Lars Magnus Hvattum

Abstract

Seagoing vessels require regular maintenance. Preventive maintenance can be planned ahead of time, and can be executed either at sea or while visiting a port. The spare parts used when performing maintenance can come from warehouses that store the parts until needed at a port, but they can also come from on-vessel storages. Some spare parts must be available on a vessel at all times, in the case that corrective maintenance is required. This chapter considers liner shipping, where each vessel follows a preplanned itinerary and a number of preventive maintenance tasks have been scheduled over time. A mathematical model is proposed that can be used to decide where to store spare parts, how many spare parts to keep in inventory, when to order spare parts from suppliers, and when and where to perform the scheduled maintenance tasks. Numerical experiments show that the model can be solved very quickly. The model can thus be used as a tool to support making decisions related to inventory management for spare parts.

Keywords: maritime transport, inventory management, maintenance, linear integer programming, mathematical programming

1. Introduction

Maritime transportation is a vital mode for national and international trade, making it one of the most important means of transportation [1]. Its historical importance has persisted over time [2]. Moreover, the exploration of maritime transportation from various research angles underscores its relevance in today's world.

Ensuring the reliability and safety of vessels is crucial in maritime transportation. Achieving this involves performing maintenance tasks that enable proper vessel functioning and prevent system failures. Effective maintenance requires a comprehensive approach with thorough long-term planning.

Maintenance planning has two key categories: Preventive and corrective maintenance [3]. Preventive maintenance occurs during system operation, aimed at preserving an item in a specific condition. On the other hand, corrective maintenance addresses system failures [4]. While preventive maintenance is important to maintenance planning, it cannot completely eliminate failures. Therefore, corrective maintenance has significant importance in maintenance planning as well [5].

IntechOpen

When undertaking maintenance tasks, certain prerequisites come into play. Among the crucial aspects is the accessibility of spare parts. In the setting of maritime transportation, spare parts can be stored on board vessels or in warehouses, or they can be delivered directly from suppliers to a vessel visiting a port.

Addressing spare part inventory management becomes crucial when the option to store these components for future maintenance arises. Efficient spare part inventory management aligns with their availability, directly influencing vessel effectiveness and performance [6].

The demand for spare parts changes over time due to factors such as maintenance requirements [7]. To mitigate this variability, maintaining a stock of spare parts provides security [8]. Moreover, the number of items in inventory can be huge due to different demands [9].

While minimizing the number of spare parts is feasible, it should not compromise vessel availability [10]. Insufficient availability of spare parts raises the risk of vessel failures [11]. On the other hand, an excess of spare parts escalates holding costs [12].

In this research, spare part inventory management in maritime transportation is explored. The motivation is a liner shipping company, where vessels follow planned, cyclic itineraries, and where each vessel has a number of planned preventive maintenance tasks. A known number of spare parts are used when executing the maintenance tasks. A single type of spare part is considered, which is ordered from a supplier that can deliver the spare parts either to warehouses located near ports or directly to a vessel when visiting a port. The goal is to determine when to order from the supplier, how much inventory of spare parts to hold, and the exact timing of performing maintenance, so that the total costs are minimized. To address the problem, a new mathematical model is introduced.

This chapter continues by first giving a review of the related research literature. Then, the problem at hand is described. Subsequently, a mathematical model is formulated for the problem. This is followed by a discussion of computational experiments, where the model is tested to illustrate its capabilities to function as a decision support tool. Finally, concluding remarks are provided, including a discussion of directions for future research.

2. Literature review

Maintenance of vessels and spare part inventory management are the main focus of the following literature review. The literature underscores the critical role of maritime maintenance in strategic decision-making processes. Within these discussions, preventive and corrective maintenance are considered as two fundamental approaches highlighted in the literature. Additionally, spare part management is considered a critical issue in effective maintenance planning.

This research specifically addresses the connection between maintenance tasks within the maritime industry and the management of spare part inventories. The literature review is designed to establish this connection, in particular from the perspective of applying optimization models for decision support.

2.1 Importance of maritime maintenance

Maintenance can increase the efficiency of a vessel and the reliability of its services. Moreover, decisions related to maritime maintenance have a significant effect on other players in a vessel's operation. Maintenance plays a crucial role in a ship's total life cycle expenses, emphasizing the need for effective management. This cost encompasses various elements, spanning from spare parts to personnel expenses [13]. Moreover, the correlation between a vessel's age and maintenance cost shows the significance of maintenance practices [14]. Turan et al. offered a model considering costs across a vessel's life cycle, showing maintenance expenses accounting for approximately one-third of the total life cycle cost [15]. Their work underscores the importance of optimizing maintenance costs and their far-reaching impact on a vessel's operations.

2.2 Preventive maintenance

In maritime operations, preventive maintenance is essential, planned to avoid expected breakdowns. Challenges include securing skilled staff, managing spare parts, and overseeing inventories in multiple locations. Pillay et al. analyzed operational delays, a key factor in assessing preventive maintenance [16]. Optimizing these tasks and schedules assists decision-makers in considering cost-effectiveness [17].

Sustainability influences maintenance strategies. Franciosi et al. explored sustainability in maintenance models, emphasizing emission reduction [18]. Liu et al. focused on environmental optimization in maintenance, particularly on engine performance's impact on greenhouse gas emissions [19].

2.3 Spare part management

The maritime industry relies heavily on spare parts for efficient vessel maintenance, emphasizing the importance of their reliability and availability. These components can be strategically stored in warehouses or on board vessels, with the capability of direct restocking by suppliers as needed. Proper spare part management is essential for enhancing vessel reliability, ensuring safety, and reducing maintenance costs. Rinaldi et al. conducted research demonstrating a correlation between the total cost and the maximum number of spare parts storages, underscoring the significance of spare part management in the broader context of asset management [20].

The delivery of spare parts has a critical role in effective spare part management, influencing vessel reliability and delivery costs. Wagner et al. addressed logistical concerns, presenting a strategic framework applicable to diverse businesses and scenarios [7]. This research highlighted the potential of logistical-based planning in spare part management. Vukić et al. proposed an optimal spare parts delivery method for vessels, considering various transportation scenarios and establishing a routing solution [21]. Their study emphasized the significance of spare part availability for shipping companies and underscored the interconnectedness of different supply chain elements in ensuring accessibility.

The demand for spare parts, varying in quantity and type, requires forecasting based on available information. Wang and Syntetos introduced an approach to forecast spare parts demand, emphasizing the critical role of demand information in optimizing spare part inventory management [22]. Van der Auweraer and Boute considered maintenance plans and system failure behavior in forecasting spare parts demand, highlighting the important role of information in accurate predictions [23].

Managing a vast number of spare parts in the maritime industry introduces complexity to spare part management. Sheikh-Zadeh et al. proposed a grouping model for spare parts management, underscoring the importance of categorization in enhancing inventory management efficiency [24]. Additionally, Cakmak and Guney focused on spare part classification as a means to reduce inventory management time, addressing the challenge posed by the industry's extensive spare parts inventory [25].

Effectively managing spare parts inventory in maritime operations involves balancing the costs of excess storage against the risks of stock-outs. Turrini and Meissner emphasized the high penalties associated with stock-out situations, highlighting the need for a strategic approach [26]. Zheng et al. proposed a solution by integrating ordering and maintenance optimization to strike a balance between storage and stock-out costs [27]. Anglou et al. contributed by presenting an approach for maritime companies to improve order management, considering factors like supplier selection, which significantly influences cost estimates and overall effectiveness [28]. This research underscored the critical role of strategic supplier decisions in reducing costs, ensuring quality, and optimizing spare parts inventory in maritime settings.

Nenni and Schiraldi emphasized the importance of optimized inventory management to address the costly storage of spare parts in maritime operations [29]. On a technological front, Kostidi et al. presented the potential of additive manufacturing to improve spare part availability and reliability while minimizing storage space and costs [30].

Sleptchenko et al. emphasized the importance of strategic planning in repairing spare parts to reduce time and costs [31]. Sustainability concerns were tackled by Driessen et al. [32] and Pater and Mitici [33], showcasing the efficiency gains and cost reductions achievable through effective spare part management. Huiskonen's work highlighted the managerial challenge of selecting spare parts, emphasizing the role of logistical considerations in this process [34].

2.4 Maintenance optimization

Optimization of maritime maintenance tasks, integral to a vessel's lifecycle, demands a comprehensive examination of its interplay with spare parts management. Notably, the scheduling of maintenance tasks emerges as a critical facet of this optimization, with Kian et al. proposing a mathematical model that strategically considers location, timing, and predictive accuracy, emphasizing its role in cost-effective and timely spare parts management [35].

The relationship between maintenance and spare part management is an important point. Wang introduced a mathematical optimization model that considered the connection between these two components [36]. Also, research by Abderrahmane et al. focused on optimal maintenance frequencies within a finite time horizon, recognizing the complex relationship between maintenance and spare part management [37]. Similarly, Eruguz et al. focused on the impact of spare part management on maintenance tasks, emphasizing the comprehensive nature of their integration and its consequential effect on operational efficiency [38].

The combination of maintenance and spare part management enhances the reliability of a system [39]. This provides a practical solution but also poses challenges for managers dealing with the complexity of the issue. Furthermore, the incorporation of uncertainty into maritime maintenance optimization, as explored by Manea et al. [40], underscored the adaptability of strategies when faced with uncertainty in costs and labor resource limitations.

In this situation, the comprehensive strategy of optimizing maintenance, along with managing spare parts inventory, emphasizes the crucial importance of maintenance planning. Basten and Ryan explore the consequences of delayed maintenance on spare parts inventory management, considering costs and highlighting the interconnected relationship of these aspects in the operational environment of the maritime industry [41].

2.5 Spare part inventory optimization

The main goal of spare part inventory optimization is to find the best stock level for spare parts, closely linked to maintenance needs [42]. Spare part inventories within the maritime industry, spanning vessels to warehouses, enhance overall availability, as demonstrated by Zhu and Zhou, who highlighted the use of multiple inventories with varying spare part levels across industries [43].

Louit et al. contributed three key optimization criteria for inventory management: Minimizing costs, maximizing availability, and achieving predefined reliability, accounting for assumed demand rates [44]. Zhang et al. structure their inventory optimization model around minimizing spare part inventory levels and total costs, encompassing transportation, inventory holding, and time-related expenses [45]. Transportation costs are a significant consideration, as demonstrated by Levner et al., who emphasized the impact of transportation costs on spare part inventory optimization [46].

Considering optimizing spare part inventory management with maintenance, Eruguz et al. proposed an integrated approach that minimizes delivery, replacement, and inventory holding costs [11]. Jiang et al. [47] and Zhang et al. [48] extended this integration by considering preventive intervals, inspection, and maximum inventory levels as decision variables.

The large quantity of spare parts on ships shows the importance of categorization to simplify the optimization process. Ben Hmida et al. [49] classified critical spare parts to decrease downtime costs in failure scenarios, while Muniz et al. [50] focused on minimizing inventory levels while maximizing criticality.

Demand dynamics, influenced by maintenance tasks and system failures, are critical considerations in spare part inventory management [51]. Zhu et al. focused on demand forecasting driven by planned maintenance, emphasizing the significance of integrating maintenance planning into spare part demand analysis [52].

3. Problem description

Liner shipping companies define routes and schedules for their vessels traveling between different ports. During each port visit, ships have the capability to pick up or deliver goods or passengers and receive various services. The routes are predetermined with fixed schedules. **Figure 1** provides an illustration of a liner service with 13 ports, a fixed 42-day route duration, and the distances and durations between ports are known. The fixed nature of the routes and port visits allows for advanced planning and scheduling.

For each vessel, preventive maintenance tasks are planned ahead of time. Adequate spare parts are crucial for maintenance, emphasizing the importance of knowing the location and quantity of each stored spare part for accessibility. Both the itineraries of the vessels and their maintenance plans are considered as inputs. There is some flexibility in determining the exact timing of performing the preventive maintenance. However, if a task is performed later than initially scheduled, it increases the risk of needing more expensive corrective maintenance, and if a task is performed

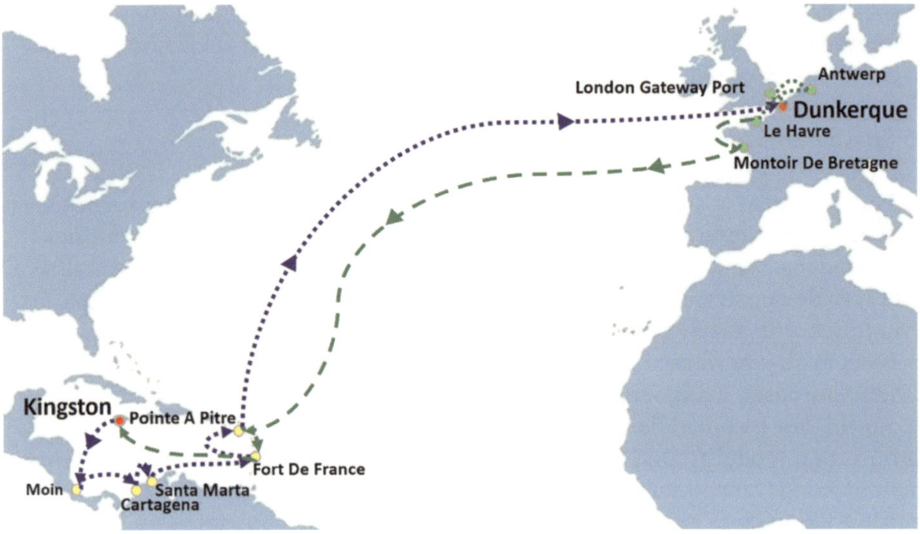

Figure 1.
An example of a liner shipping service [53].

earlier than initially scheduled, it means that the next preventive maintenance may be needed earlier than it would otherwise be.

The planner must determine the optimal number of spare parts on vessels and at warehouses, focusing on planned (preventive) maintenance and associated spare parts inventory management. It is crucial to strike a balance between having enough spare parts to execute the planned maintenance and minimize the number of spare parts on a vessel, and in warehouses, together with the associated storage costs. There are also space limitations for storage, and the vessel stops at different ports with varying access to warehouses. In the problem at hand, only one type of spare parts is considered.

The problem scenario involves the requirement for spare parts across different maintenance tasks for various vessels, each with distinct spare part needs and maintenance task due dates. Spare parts may be stored on vessels, in warehouses, or ordered and delivered directly from suppliers, each with associated storage costs and capacity constraints.

Maintenance tasks can occur at sea or in ports, with spare parts sourced directly from vessel or warehouse storage, suppliers, or a combination of these. The initial inventory levels on vessels and in warehouses are predefined, influencing the required number of stored spare parts. Known values for inventory costs, restocking costs, and maintenance task costs, varying across vessels and ports, contribute to the decision-making process.

The timing of maintenance task execution is critical, impacting the cost-effectiveness of operations. The consideration of planned maintenance tasks on their due dates aims to prevent potential future corrective maintenance needs. Decisions revolve around determining the number of stored spare parts and restocking strategies, with **Table 1** providing a comprehensive overview of the problem's key information and decision points.

Figure 2 illustrates the flow of spare parts. The supplier can deliver spare parts directly to a vessel, when the vessel is in a port, at a high cost. Alternatively, the spare

Available information	Decisions
Planning horizon	Inventory levels on vessels
Lists of ports, vessels, and maintenance plans	Inventory levels at warehouses
Current locations of vessels	Restocking decisions for vessels
Number of spare parts per maintenance	Restocking decisions for warehouses
Initial inventory levels	When to perform maintenance
Inventory holding costs	Where to perform maintenance
Restocking costs	Source of spare parts for maintenance
Maintenance costs	
Upper and lower inventory limits	

Table 1.
Overview of available information (left) and the different decisions required (right).

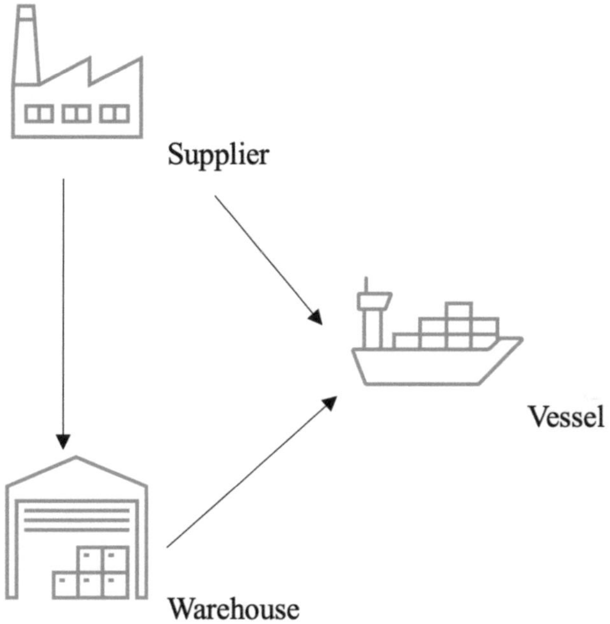

Figure 2.
Spare parts flow.

parts can be shipped to a warehouse, and then from the warehouse to the vessel when visiting a nearby port.

4. Methodology

The spare part inventory management problem described above is modeled using mathematical programming. This section introduces the notation used in the mathematical model, followed by a detailed explanation of the mathematical model.

4.1 Notation

This section introduces the sets, the parameters, and then the variables used in the mathematical model. The planning horizon is divided into equal periods, so that the whole planning horizon consists of the periods in $T = \{1, 2, \dots, H\}$. The set of ports which a vessel can visit is denoted by P. The set of vessels is V, and the set of maintenance tasks to be done on vessel $v \in V$ is dented by M_v. The model has the following parameters:

H: The number of time periods in the planning horizon.

I_{0v}^{V}: The initial inventory level on board vessel $v \in V$.

I_{0p}^{W}: The initial inventory level on board vessel $v \in V$.

L_{tpv}: Indicator for whether or not a given vessel $v \in V$ is located at a given port $p \in P$ at a given time period $t \in T$.

N_{vm}: Number of spare parts needed for maintenance $m \in M_v$ for vessel $v \in V$.

C_v^{IV}: Inventory cost for a spare part on board vessel $v \in V$.

C_p^{IW} : Inventory cost for a spare part at a warehouse at port $p \in P$.

C_p^{RV} : Restocking cost of a spare part from a supplier directly to a vessel when the vessel visits port $p \in P$.

C_p^{RW} : Restocking cost of a spare part from a supplier to a warehouse at port $p \in P$.

C_p^{RWV} : Restocking cost of a spare part from a warehouse at port $p \in P$ to a vessel visiting the port.

C_v^{MV}: Maintenance cost for tasks done by the crew of vessel $v \in V$ at sea.

C_p^{MP}: Maintenance cost for tasks done at port $p \in P$.

C_{vmt}^{MT}: Penalty cost for performing maintenance $m \in M_v$ for vessel $v \in V$ in period $t \in T$.

I_v^{maxV}: Maximum number of spare parts stored on vessel $v \in V$.

I_p^{maxW}: Maximum number of spare parts stored in the warehouse at port $p \in P$.

I_v^{minV}: Minimum number of spare parts which should be available on board vessel $v \in V$.

I_p^{minW}: The minimum number of spare parts which should be stored in the warehouse at port $p \in P$.

The decision variables considered in the mathematical model are as follows:

I_{tv}^{V}: The inventory level of vessel $v \in V$ in period $t \in T$.

I_{tp}^{w}: The inventory level of a warehouse in port $p \in P$ in period, $t \in T$.

X_{tpv}^{RV}: Number of spare parts to be restocked at period $t \in T$ from a supplier directly to vessel $v \in V$ when the vessel visits port $p \in P$.

Y_{tpv}^{RV}: Binary variable equal to 1 if and only if a spare part is restocked at period $t \in T$ from a supplier directly to vessel $v \in V$ when the vessel visits port $p \in P$.

X_{tp}^{RW}: Number of spare parts to be restocked at time period $t \in T$ from a supplier to a warehouse at port $p \in P$.

Y_{tp}^{RW}: Binary variable equal to 1 if and only if a spare part is restocked at period $t \in T$ from a supplier to a warehouse at port $p \in P$.

X_{tpv}^{RWV}: Number of spare parts to be restocked at period $t \in T$ from a warehouse at port $p \in P$ to vessel $v \in V$ visiting the port.

Y^{RWV}_{tpv}: Binary variable equal to 1 if and only if a spare part is restocked in period $t \in T$ from a warehouse at port $p \in P$ to vessel $v \in V$ visiting the port.

X^{MV}_{vmt}: Number of spare parts coming from storage on board vessel $v \in V$ used for maintenance task $m \in M_v$ at period $t \in T$.

Y^{MV}_{vmt}: Binary variable equal to 1 if and only if maintenance task $m \in M_v$on vessel $v \in V$ is performed in period $t \in T$ at sea using the vessel's own crew.

X^{MW}_{vpmt}: Number of a spare parts coming from a warehouse at port $p \in P$ to be used for maintenance task $m \in M_v$, $v \in V$, at period $t \in T$.

Y^{MP}_{vpmt}: Binary variable equal to 1 if and only if maintenance task $m \in M_v$ on vessel $v \in V$ is performed by the personnel of port $p \in P$ when the vessel visits the port at time period $t \in T$.

4.2 Mathematical model

Using the mathematical notation defined above, the following mathematical programming model is proposed.

$$\begin{aligned} \min \sum_{t \in T} \sum_{v \in V} I^V_{tv} C^{IV}_v + \sum_{t \in T} \sum_{p \in P} I^w_{tp} C^{Iw}_p + \sum_{t \in T} \sum_{p \in P} \sum_{v \in V} X^{RV}_{tpv} C^{RV}_p + \sum_{t \in T} \sum_{p \in P} X^{RW}_{tp} C^{RW}_p \\ + \sum_{t \in T} \sum_{p \in P} \sum_{v \in V} X^{RWV}_{tpv} C^{RWV}_p + \sum_{v \in V} \sum_{m \in M_v} \sum_{t \in T} Y^{MV}_{vmt} C^{MV}_v \\ + \sum_{v \in V} \sum_{p \in P} \sum_{m \in M_v} \sum_{t \in T} Y^{MP}_{vpmt} C^{MP}_p + \sum_{v \in V} \sum_{m \in M_v} \sum_{t \in T} Y^{MV}_{vmt} C^{MT}_{vmt} \\ + \sum_{v \in V} \sum_{p \in P} \sum_{m \in M_v} \sum_{t \in T} Y^{MP}_{vpmt} C^{MT}_{vmt} \end{aligned} \tag{1}$$

The objective function in Eq. (1) consists of inventory costs for vessels and warehouses; restocking costs between suppliers, vessels, and warehouses; and maintenance costs that depend on whether the maintenance is performed at sea or at port, as well as the time period in which the maintenance is performed. The constraints of the model follow:

$$I^V_{tv} \le I^{maxV}_v \quad t \in T, v \in V \tag{2}$$

Constraints (2) make sure that the number of spare parts on board of a vessel never exceeds the considered upper limit for this spare part.

$$I^w_{tp} \le I^{maxW}_p \quad t \in T, p \in P \tag{3}$$

The number of spare parts at each warehouse at a port never exceeds the corresponding capacity at the warehouse. This is modeled using constraints (3), which can also be used to indicate that a warehouse is not available, by setting the right hand side to zero.

$$I^V_{tv} \ge I^{minV}_v \quad t \in T, v \in V \tag{4}$$

$$I^w_{tp} \ge I^{minW}_p \quad t \in T, p \in P \tag{5}$$

There are also lower limits for the number of spare parts on board of vessels and in warehouses, as indicated in constraints (4) and (5).

$$I_{tp}^{W} = I_{tp}^{W} + X_{tp}^{RW} - \sum_{v \in V} X_{tpv}^{RWV} - \sum_{v \in V} \sum_{m \in M_v} X_{vpmt}^{MW} \quad t \in T, p \in P \tag{6}$$

The inventory level of a spare part at each warehouse at a port during each period is related to the inventory level of the previous time period. Also, each warehouse can receive spare parts from a supplier in each period. Moreover, spare parts can be sent from a warehouse to a vessel for a maintenance task, and to restock the inventory on board of the vessel. All these considerations are represented in constraints (6).

$$X_{tp}^{RW} \leq I_{p}^{maxW} Y_{tp}^{RW} \quad t \in T, p \in P \tag{7}$$

Constraints (7) make sure that the binary variable Y_{tp}^{RW} takes a value of 1 if any units are sent from a supplier to the warehouse at port p in time period t. It also limits the quantity that can be sent in a single period, which is set equal to the capacity of the warehouse.

$$X_{vpmt}^{MW} \leq I_{p}^{maxW} Y_{vpmt}^{MW} \quad t \in T, v \in V, p \in P, m \in M_v \tag{8}$$

As the previous constraints, constraints (8) are used to set binary variables to 1 to indicate the presence of a flow of spare parts. In this case, the flow is from a warehouse to a vessel where the parts are used immediately in a maintenance operation.

$$Y_{vpmt}^{MW} \leq L_{tpv} \quad t \in T, v \in V, p \in P, m \in M_v \tag{9}$$

A spare part can only be picked from a warehouse for a maintenance task on a vessel when the vessel visits a port that has access to the considered warehouse, as enforced by constraints (9).

$$I_{tv}^{V} = I_{tv}^{V} + \sum_{p \in P} X_{tpv}^{RV} + \sum_{p \in P} X_{tpv}^{RWV} - \sum_{m \in Mv} X_{vmt}^{MV} \quad t \in T, v \in V \tag{10}$$

Constraints (10) define the level of inventory of each vessel at the end of each period, which has a direct relationship with the previous inventory level. Each vessel can receive spare parts directly from a supplier or a warehouse when it visits a port that has access to them. Moreover, spare parts on board of a vessel can be directly used for a maintenance task.

$$Y_{tpv}^{RV} \leq L_{tpv} \qquad t \in T, p \in P, v \in V \tag{11}$$

A vessel can only receive spare parts from a supplier when it visits a port that has access to the supplier, as enforced by constraints (11). Furthermore, there is a possibility for a vessel of receiving spare parts from a warehouse when it visits a port that has access to the warehouse, as given by constraints (12).

$$Y_{tpv}^{RWV} \leq L_{tpv} \qquad t \in T, p \in P, v \in V \tag{12}$$

$$X_{tpv}^{RV} \leq I_{v}^{maxV} Y_{tpv}^{RV} \qquad t \in T, p \in P, v \in V \tag{13}$$

When spare parts are sent to a vessel from a supplier, a corresponding binary variable Y^{RV}_{tpv} must be forced to 1, so that the related costs can be calculated in the objective function. This is ensured in constraints (13). A similar connection is made in constraints (14), regarding spare parts that are sent from a warehouse to a vessel.

$$X^{RWV}_{tpv} \leq I^{maxV}_{v} Y^{RWV}_{tpv} \quad t \in T, p \in P, v \in V \tag{14}$$

$$X^{MV}_{vmt} \leq I^{maxV}_{v} Y^{MV}_{vmt} \quad t \in T, v \in V, m \in M_v \tag{15}$$

It is possible to get the spare parts from storage areas on board of a vessel to be used directly for a maintenance task. Then, the corresponding binary variable Y^{MV}_{vmt} should be set to 1, as ensured by constraints (15).

$$\sum_{t \in T} Y^{MV}_{vmt} + \sum_{t \in T} \sum_{p \in P} Y^{MP}_{vpmt} = 1 \quad v \in V, m \in M_v \tag{16}$$

A maintenance task can either be done by a vessel's crew and their facilities, or by a port's personnel and their facilities when the vessel visits the port. Constraints (16) make sure that one of these options is selected for every maintenance task.

$$\sum_{p \in P} X^{MW}_{vpmt} + X^{MV}_{vmt} \geq N_{vm} \quad v \in V, m \in M_v, t \in T \tag{17}$$

For each planned maintenance task, the number of spare parts used must be sufficient. Constraints (17) make sure that this is the case after considering that the spare parts can be taken either directly from the vessel, or from a nearby warehouse.

5. Computational experiments

To evaluate the potential to use the mathematical model presented in the previous section for decision support in spare part management, a series of computational experiments are conducted. These involve solving the model for different instances, which are divided into two groups. The difference between the groups lies in the planning horizon, which is either 1 year, with 365 time periods corresponding to days, or 3 years, with 156 time periods corresponding to weeks.

Within each group of instances, a base case is defined using artificial data. This instance considers 10 vessels, 20 ports, and 2 warehouses. Each vessel has three planned maintenances in the one-year instances and six maintenances in the three-year instances. Each maintenance requires one spare part, and the capacity of each vessel is five items, whereas the capacity of each warehouse is ten items. The cost of sending spare parts directly from the supplier to a vessel is twice the cost of sending the parts to a warehouse, which again is ten times the cost of restocking a vessel using the nearby warehouse. The initial inventories are assumed to be empty, corresponding to the minimum allowed inventory levels.

Starting from the base instances, further instances are considered by varying one aspect of the underlying base instance. In particular, we investigate the effect of the following parameters:

- Number of ports with access to warehouses (increased to three)

- Number of moving vessels (reduced to nine)
- Number of maintenance tasks for each vessel (reduced by one-third)
- Number of required spare parts for each maintenance task (increased to two)
- Initial inventory level on vessels (increased by one)
- Initial inventory level at warehouses (increased by one)
- Ratio of inventory to ordering cost (halved)
- Ratio of maintenance cost by vessel's crew and facilities to maintenance cost by port's personnel and facilities (increased from two to three)
- Minimum inventory level on vessels (increased by one)

The mathematical model was implemented in AMPL and solved using the CPLEX 20.1.0 solver on a computer running macOS Ventura equipped with 16 GB RAM and an Apple M2 processor. The CPLEX provides a mixed integer programming solver based on the branch-and-bound method, applying the simplex method to solve linear programming relaxations, and using advanced heuristics and cut-generation.

The generated instances can be solved to optimality quickly, with running times ranging from 1.7 to 15.6 seconds. **Table 2** shows the individual running times as well as the objective function values for each instance solved.

Examining the instances with a one-year planning horizon, the frequency of direct spare part orders from suppliers to vessels is greatly influenced by factors such as the quantity of needed spare parts and the related ordering costs. The solutions show a sensitivity to ordering costs, avoiding direct orders when costs are increased. In contrast, with a three-year planning horizon, the ordering patterns are seen to vary based

		One-year horizon		Three-year horizon	
Instance	**Parameter varied**	**Obj. fun**	**Seconds**	**Obj. fun**	**Seconds**
Base	None	8954.0	3.1	15585.0	1.9
1	#Ports	8834.0	3.2	13830.0	1.8
2	#Vessels	8194.0	2.9	14055.0	1.7
3	#Tasks	5987.0	2.1	10391.0	1.3
4	#Parts per task	12538.0	12.2	22087.8	8.4
5	Vessel inventory	9459.0	15.6	16802.0	6.4
6	Warehouse inventory	8765.8	3.3	15803.4	1.8
7	Inventory/ordering cost	11970.0	3.2	21685.0	2.3
8	Maintenance cost	9854.0	3.0	16185.0	1.8
9	Minimum inventory	12604.0	3.3	26508.0	1.8

Table 2.
Numerical results from solving twenty test instances.

on factors such as the number of needed spare parts and the availability of warehouses. Additionally, the analysis indicates that there are more deliveries to warehouses when there are more warehouses available. An increase in the initial inventory levels on vessels and warehouses reduces the necessity for placing orders with the supplier.

Based on the analysis, the highest overall number of maintenance tasks carried out by the crews of the vessels at sea is observed in the one-year planning horizon when there is an initial inventory of spare parts on board of vessels. On the other hand, in the instance where the need for maintenance tasks is lower, the number of maintenance tasks performed at sea is the minimum among all instances.

In the three-year planning horizon, the pattern of the greatest number of maintenance tasks performed at sea aligns with the observations for the one-year planning horizon. Conversely, the instance with the highest number of warehouses has the lowest number of maintenance tasks performed at sea.

According to the findings, the instances with a non-zero initial inventory level for vessels exhibit the highest total number of time periods where vessels maintain an inventory of spare parts exceeding the minimum level, observed in both the one-year and three-year planning horizons. This shows that it is unnecessary to keep more than the minimum inventory level on board of vessels, as in the case studied the vessels are visiting ports with warehouses relatively frequently.

In summary, the mathematical model provided can be solved very efficiently, even for long planning horizons. This allows decision makers to test different scenarios, to see how different situations may affect the need for keeping different amounts of spare parts in inventory.

6. Conclusion

Transportation is an essential part of our lives. Our daily plans are affected directly and indirectly by transportation which exists in different forms. Among transportation modes, maritime transportation plays an important role which expands different areas of research around it. The vessels used in maritime transportation require regular maintenance, which involves the use of spare parts, some of which are critical to the operation of the vessels.

Inventory management of spare parts for vessels has not been studied in depth in the research literature [54]. The planning involved can be complex, as spare parts can be stored both on board vessels and in warehouses located near ports. To minimize costs, the planners must balance the need for having spare parts available at the times of performing planned preventive maintenance tasks, while balancing the costs of ordering from suppliers and the cost of holding inventory.

This chapter discusses one particular optimization problem arising under certain assumptions. The setting is taken from liner shipping, where each vessel has a predetermined itinerary that it follows in a cyclic manner. This means that decisions regarding routing and scheduling have already been fixed. Each vessel also has a set of planned maintenance tasks, but with some flexibility in deciding the exact timing of performing each of them.

The contribution of this chapter is to propose a mathematical programming model for the problem considered and to show that the model is solvable using commercially available software. Several artificially generated instances are solved, which gives some insights into how the structure of the solution may change depending on the cost structure and other aspects of the particular instance.

This research defined artificial instances to test the model to be close to a real-world setting. The model and the presented results show the potential of using optimization for spare part inventory management in maritime maintenance.

However, there are many directions for future developments. First, the model only considers preventive maintenance tasks. In reality, the need for corrective maintenance tasks arises dynamically and must be handled. While the current model can be adopted by enforcing a safety stock (setting the lower limit of inventory at vessels and warehouses to equal the safety stock), it may be better to explicitly model the uncertainty, and to formulate this as a dynamic optimization problem.

Second, in this context, liner shipping is an easy mode, as the routes and schedules of each vessel are planned well in advance. For tramp shipping and industrial shipping, the routes are much more dynamic. In those areas, it may be more pertinent to plan the timing of preventive maintenance simultaneously while planning the routes and schedules of vessels. This increases the difficulty of the planning problem significantly.

Third, the problem considered in this chapter is a tactical problem, and there are strategic decisions that play an important role. This concerns for example the location of warehouses, the capacity of warehouses, and the selection of suppliers.

Fourth, the considered problem takes into account only a single type of spare parts. In reality, each maintenance task requires a large number of different spare parts, of different complexity. Thus, the timing of the maintenance tasks and the ordering from suppliers may need to be coordinated. This is particularly true if a single supplier can provide several different types of spare parts, and the ordering costs are nonlinear in the number of spare parts ordered.

In practice, future research is required to see how maritime transportation companies can apply planning tools as presented here. In the case of incorporating corrective maintenance, these companies require good data collection abilities to ensure good managerial decisions.

As environmental concerns increase every day, there is a need to extend the knowledge in the maritime industry in a way of respecting this issue. This can be a good way of thinking for future research by including fuel consumption and the environmental impact of spare part inventory management in the presented model.

As technology arises by exploring different areas of knowledge, it is necessary to adopt new technologies and use them in the best way of respecting the environment and society. A fast-growing technology is additive manufacturing. This can be also a good way of extending the research in the future to make a connection between additive manufacturing and spare part inventory management. The circularity of a business can be also a good way to expand the presented model for future research which can be done by working on the reverse logistics of unused or damaged spare parts.

Acknowledgements

The authors thank the editor for valuable comments that helped to improve this chapter.

Conflict of interest

The authors declare no conflict of interest.

Author details

Arameh Bisadi, Amir Zare and Lars Magnus Hvattum*
Molde University College, Molde, Norway

*Address all correspondence to: lars.m.hvattum@himolde.no

References

[1] United Nations Conference on Trade and Development. Review of Maritime Transport 2022 [Internet]. UNCTAD; 2022. Available from: https://unctad.org/rmt2022

[2] Rodrigue J-P. The Geography of Transport Systems. 5th ed. London: Routledge; 2020. DOI: 10.4324/9780429346323

[3] Zhao J, Yang L. A bi-objective model for vessel emergency maintenance under a condition-based maintenance strategy. SIMULATION. 2017;**94**(7):609-624. DOI: 10.1177/0037549717741973

[4] Wang H. A survey of maintenance policies of deteriorating systems. European Journal of Operational Research. 2002;**139**(3):469-489. DOI: 10.1016/s0377-2217(01)00197-7

[5] Zhang S, Huang K, Yuan Y. Spare parts inventory management: A literature review. Sustainability. 2021; **13**(5):2460. DOI: 10.3390/su13052460

[6] Barabadi A, Barabady J, Markeset T. Application of reliability models with covariates in spare part prediction and optimization – A case study. Reliability Engineering & System Safety. 2014;**123**:1-7. DOI: 10.1016/j.ress.2013.09.012

[7] Wagner SM, Jönke R, Eisingerich AB. A strategic framework for spare parts logistics. California Management Review. 2012;**54**(4):69-92. DOI: 10.1525/cmr.2012.54.4.69

[8] Şahin M, Eldemir F, Turkyilmaz A. Inventory cost minimization of spare parts in aviation industry. Transportation Research Procedia. 2021; **59**:29-37. DOI: 10.1016/j.trpro.2021.11.094

[9] Altay Guvenir H, Erel E. Multicriteria inventory classification using a genetic algorithm. European Journal of Operational Research. 1998;**105**(1):29-37. DOI: 10.1016/s0377-2217(97)00039-8

[10] Basten RJI, van Houtum GJ. System-oriented inventory models for spare parts. Surveys in Operations Research and Management Science. 2014;**19**(1):34-55. DOI: 10.1016/j.sorms.2014.05.002

[11] Eruguz AS, Tan T, van Houtum GJ. A survey of maintenance and service logistics management: Classification and research agenda from a maritime sector perspective. Computers & Operations Research. 2017;**85**:184-205. DOI: 10.1080/24725854.2017.1312037

[12] Lin B, Wang J, Wang H, Wang Z, Li J, Lin R, et al. Inventory-transportation integrated optimization for maintenance spare parts of high-speed trains. PLoS One. 2017;12(5):e0176961. DOI: 10.1371/journal.pone.0176961

[13] Schuh P, Schneider D, Funke L, Tracht K. Cost-optimal spare parts inventory planning for wind energy systems. Logistics Research. 2015;**8**(1):8. Article 4. DOI: 10.1007/s12159-015-0122-7

[14] Kimera D, Nangolo FN. Maintenance practices and parameters for marine mechanical systems: A review. Journal of Quality in Maintenance Engineering. 2020;**26**(3):459-488. DOI: 10.1108/jqme-03-2019-0026

[15] Turan O, Ölçer A, Lazakis I, Rigo P, Caprace JD. Maintenance/repair and production-oriented life cycle cost/earning model for ship structural

optimisation during conceptual design stage. Ships and Offshore Structures. 2009;**4**(2):107-125. DOI: 10.1080/17445300802564220

[16] Pillay A, Wang J, Wall AD, Ruxton T. A maintenance study of fishing vessel equipment using delay-time analysis. Journal of Quality in Maintenance Engineering. 2001;7(2):118-128. DOI: 10.1108/13552510110397421

[17] Lanza G, Niggeschmidt S, Werner P. Optimization of preventive maintenance and spare part provision for machine tools based on variable operational conditions. CIRP Annals. 2009;**58**(1):429-432. DOI: 10.1016/j.cirp.2009.03.030

[18] Franciosi C, Lambiase A, Miranda S. Sustainable maintenance: A periodic preventive maintenance model with sustainable spare parts management. IFAC-PapersOnLine. 2017;**50**(1): 13692-13697. DOI: 10.1016/j.ifacol.2017.08.2536

[19] Liu S, Chen H, Shang B, Papanikolaou A. Supporting predictive maintenance of a ship by analysis of onboard measurements. Journal of Marine Science and Engineering. 2022;**10**(2):215. DOI: 10.3390/jmse10020215

[20] Rinaldi M, Fera M, Macchiaroli R, Bottani E. A new procedure for spare parts inventory management in ETO production: A case study. Procedia Computer Science. 2023;**217**:376-385. DOI: 10.1016/j.procs.2022.12.233

[21] Vukić L, Stazić L, Pijaca M, Peronja I. Modelling the optimal delivery of spare parts to vessels: Comparison of three different scenarios. Cogent Engineering. 2021;**8**(1):19. Article no 1920565. DOI: 10.1080/23311916.2021.1920565

[22] Wang W, Syntetos AA. Spare parts demand: Linking forecasting to equipment maintenance. Transportation Research Part E: Logistics and Transportation Review. 2011;**47**(6): 1194-1209. DOI: 10.1016/j.tre.2011.04.008

[23] Van der Auweraer S, Boute R. Forecasting spare part demand using service maintenance information. International Journal of Production Economics. 2019;**213**:138-149. DOI: 10.1016/j.ijpe.2019.03.015

[24] Sheikh-Zadeh A, Farhangi H, Rossetti MD. Inventory grouping and sensitivity analysis in multi-echelon spare part provisioning systems. Computers & Industrial Engineering. 2020;**143**:106230. DOI: 10.1016/j.cie.2019.106230

[25] Cakmak E, Guney E. Spare parts inventory classification using neutrosophic fuzzy EDAS method in the aviation industry. Expert Systems with Applications. 2023;**224**:120008. DOI: 10.1016/j.eswa.2023.120008

[26] Turrini L, Meissner J. Spare parts inventory management: New evidence from distribution fitting. European Journal of Operational Research. 2019; **273**(1):118-130. DOI: 10.1016/j.ejor.2017.09.039

[27] Zheng M, Lin J, Xia T, Liu Y, Pan E. Joint condition-based maintenance and spare provisioning policy for a K-out-of-N system with failures during inspection intervals. European Journal of Operational Research. 2023;**308**(3): 1220-1232. DOI: 10.1016/j.ejor.2023.01.011

[28] Anglou F, Ponis S, Spanos A. A machine learning approach to enable bulk orders of critical spare-parts in the shipping industry. Journal of Industrial

Engineering and Management. 2021; **14**(3):604. DOI: 10.3926/jiem.3446

[29] Nenni M, Schiraldi M. Optimizing spare parts inventory in shipping industry. International Journal of Engineering & Technology. 2013;**5**: 3152-3157

[30] Kostidi E, Nikitakos N, Progoulakis I. Additive manufacturing and maritime spare parts: Benefits and obstacles for the end-users. Journal of Marine Science and Engineering. 2021; **9**(8):895. DOI: 10.3390/jmse9080895

[31] Sleptchenko A, van der Heijden MC, van Harten A. Using repair priorities to reduce stock investment in spare part networks. European Journal of Operational Research. 2005;**163**(3): 733-750. DOI: 10.1016/j.ejor.2004. 02.002

[32] Driessen M, Arts J, van Houtum G-J, Rustenburg JW, Huisman B. Maintenance spare parts planning and control: A framework for control and agenda for future research. Production Planning & Control. 2015:**26**(5):407-426. DOI: 10.1080/09537287.2014.907586

[33] Pater I de, Mitici M. Predictive maintenance for multi-component systems of repairables with remaining-useful-life prognostics and a limited stock of spare components. Reliability Engineering & System Safety. 2021;**214**: 107761. DOI: 10.1016/j.ress.2021.107761

[34] Huiskonen J. Maintenance spare parts logistics: Special characteristics and strategic choices. International Journal of Production Economics. 2001;**71**(1–3): 125-133. DOI: 10.1016/s0925-5273(00) 00112-2

[35] Kian R, Bektaş T, Ouelhadj D. Optimal spare parts management for vessel maintenance scheduling. Annals of Operations Research. 2018;**272**(1–2): 323-353. DOI: 10.1007/s10479-018-2907-y

[36] Wang W. A joint spare part and maintenance inspection optimisation model using the delay-time concept. Reliability Engineering & System Safety. 2011;**96**(11):1535-1541. DOI: 10.1016/ j.ress.2011.07.004

[37] Abderrahmane F, Bouslikhane S, Hajej Z, Dellagi S, Trabelsi W. An improved integrated maintenance/spare parts management for wind turbine systems with adopting switching concept. Energy Reports. 2022;**8**: 936-955. DOI: 10.1016/j.egyr.2022.07.123

[38] Eruguz AS, Tan T, van Houtum G-J. Integrated maintenance and spare part optimization for moving assets. IISE Transactions. 2017;**50**(3):230-245. DOI: 10.1080/24725854.2017.1312037

[39] Fan L, Su H, Zio E, Li Y, Zhang L, Peng S, et al. Supply reliability-driven joint optimization of maintenance and spare parts inventory in a gas pipeline system. Gas Science and Engineering. 2023;**110**:204883. DOI: 10.1016/ j.jgsce.2023.204883

[40] Manea MG, Zăgan R, Manea E. Comments regarding the use of Monte Carlo method in the analysis of maritime ship maintenance works. IOP Conference Series: Materials Science and Engineering. 2021;**1182**(1):012043. DOI: 10.1088/1757-899x/1182/1/012043

[41] Basten RJI, Ryan JK. The value of maintenance delay flexibility for improved spare parts inventory management. European Journal of Operational Research. 2019;**278**(2):646-457. DOI: 10.1016/j. ejor.2019.04.023

[42] Hu Q, Boylan JE, Chen H, Labib A. OR in spare parts management: A

review. European Journal of Operational Research. 2018;**266**(2):395-414. DOI: 10.1016/j.ejor.2017.07.058

[43] Zhu M, Zhou X. Hybrid opportunistic maintenance policy for serial-parallel multi-station manufacturing systems with spare part overlap. Reliability Engineering & System Safety. 2023;**236**:109304. DOI: 10.1016/j.ress.2023.109304

[44] Louit D, Pascual R, Banjevic D, Jardine AK. Optimization models for critical spare parts inventories—A reliability approach. Journal of the Operational Research Society. 2011; **62**(6):992-1004. DOI: 10.1057/jors.2010.49

[45] Zhang L, Deng Q, Miao B, Liu X, Shao H. Parallel service mode of production and inventory for spare part inventory optimization. Knowledge-Based Systems. 2022;**241**:108282. DOI: 10.1016/j.knosys.2022.108282

[46] Levner E, Perlman Y, Cheng TCE, Levner I. A network approach to modeling the multi-echelon spare-part inventory system with backorders and interval-valued demand. International Journal of Production Economics. 2011; **132**(1):43-51. DOI: 10.1016/j.ijpe.2011.03.004

[47] Jiang Y, Chen M, Zhou D. Joint optimization of preventive maintenance and inventory policies for multi-unit systems subject to deteriorating spare part inventory. Journal of Manufacturing Systems. 2015;**35**:191-205. DOI: 10.1016/j.jmsy.2015.01.002

[48] Zhang J, Zhao X, Song Y, Qiu Q. Joint optimization of condition-based maintenance and spares inventory for a series–parallel system with two failure modes. Computers & Industrial Engineering. 2022;**168**:108094. DOI: 10.1016/j.cie.2022.108094

[49] Ben Hmida J, Regan G, Lee J. Inventory management and maintenance in offshore vessel industry. Journal of Industrial Engineering. 2013; **2013**:1-7. DOI: 10.1155/2013/851092

[50] Muniz LR, Conceição SV, Rodrigues LF, de Freitas Almeida JF, Affonso TB. Spare parts inventory management: A new hybrid approach. The International Journal of Logistics Management. 2020;**32**(1):40-67. DOI: 10.1108/ijlm-12-2019-0361

[51] Aktepe A, Yanık E, Ersöz S. Demand forecasting application with regression and artificial intelligence methods in a construction machinery company. Journal of Intelligent Manufacturing. 2021;**32**(6):1587-1604. DOI: 10.1007/s10845-021-01737-8

[52] Zhu S, Jaarsveld W Van, Dekker R. Spare parts inventory control based on maintenance planning. Reliability Engineering & System Safety. 2020;193: 106600. DOI: 10.1016/j.ress.2019.106600

[53] CMA CGM. North Europe French West Indies (NEFWI) [Internet]. 2023. Available from: https://www.cma-cgm.com/products-services/line-services/flyer/NEFWI

[54] Bisadi A, Zare A. Maritime Maintenance Spare Part Inventory Management [Master's Degree Thesis]. Molde, Norway: Molde University College; 2023. Available from: https://hdl.handle.net/11250/3092316

Chapter 5

Data, Models, and Performance: A Comprehensive Guide to Predictive Maintenance in Industrial Settings

Kiavash Fathi and Hans Wernher van de Venn

Abstract

With the ever-growing complexity of different assets in a factory, the main focus of predictive maintenance solutions has shifted from model-based approaches to data-driven and hybrid approaches. This shift as a result highlights the importance and the inevitable impact of data, data quality, model maintenance, and model interpretability on the performance and acceptability of these predictive maintenance approaches in industry. In this chapter, the hurdles for developing effective predictive maintenance solutions for original equipment manufacturers (OEMs) and small and medium-sized enterprises (SMEs) with different levels of digitalization are introduced. Furthermore, it is discussed how to choose a suitable strategy for developing a predictive maintenance model, given the different constraints in the availability of data and the requirements of the customer.

Keywords: predictive maintenance, data quality, Industry 4.0, machine learning, trustworthy AI

1. Introduction

In the landscape of industrial operations, the concept of maintenance has undergone a profound evolution. Traditionally, maintenance strategies were predominantly reactive or scheduled, often resulting in downtime, inefficiencies, and unexpected costs. Assets would be repaired or replaced only after failure, leading to disruptions in production and compromising overall efficiency. Furthermore, scheduled maintenance, while aiming to prevent breakdowns, often resulted in unnecessary servicing of equipment that may not have required immediate attention, leading to inefficiencies and increased costs.

However, with the advent of the digital era and advancements in technology, a paradigm shift has occurred, ushering in the era of predictive maintenance. At the heart of predictive maintenance lies the integration of data, models, and performance evaluation, offering a proactive approach to asset management and optimization. By harnessing data from various sources, such as sensors, Internet of Things (IoT) devices, and historical records, predictive maintenance enables the prediction of equipment failures before they occur, allowing for timely intervention and maintenance activities. This transition from reactive or scheduled maintenance to predictive maintenance represents a significant leap forward in operational efficiency and cost-effectiveness.

The complexity of assets within modern factories has surged exponentially, accompanied by a corresponding rise in the complexity of maintenance challenges. To address these challenges, predictive maintenance solutions have transitioned from traditional model-based approaches to more data-driven and hybrid methodologies. This shift underscores the paramount importance of data and their quality, along with the ongoing maintenance and interpretability of models, in ensuring the efficacy and acceptance of predictive maintenance solutions across industries.

In this context, the availability and quality of data play a critical role in the success of predictive maintenance initiatives. The proliferation of sensors and IoT devices has enabled the generation of vast amounts of data from industrial equipment and processes. However, challenges, such as data silos, interoperability issues, and data quality concerns, can hinder the effective utilization of these data for predictive maintenance purposes. Addressing these challenges requires robust data management strategies, including data integration, cleansing, and preprocessing, to ensure the reliability and accuracy of predictive maintenance models.

Moreover, the concept of the Asset Administration Shell (AAS) in the context of predictive maintenance deserves attention. The AAS, as defined in the Reference Architecture Model Industrie 4.0 (RAMI4.0), offers a standardized representation of assets and their associated data. This framework facilitates interoperability and data exchange across heterogeneous systems, ensuring seamless integration and communication within industrial environments. By adopting the AAS framework, organizations can enhance the transparency, accessibility, and integrity of asset-related data, thereby enabling more effective predictive maintenance strategies.

As predictive maintenance solutions evolve to meet the diverse needs of original equipment manufacturers (OEMs) and small and medium-sized enterprises (SMEs) operating at varying levels of digitalization, it becomes essential to consider the specific requirements and constraints of each context. Customer requirements, including cost considerations, operational priorities, and regulatory compliance, must be carefully evaluated when developing predictive maintenance solutions. Balancing these requirements with the constraints imposed by data availability and technological capabilities is crucial in selecting an optimal strategy for predictive maintenance model development.

Throughout this chapter, we will navigate through the complexities and nuances of predictive maintenance in industrial settings, exploring challenges, opportunities, and best practices for leveraging data, models, and performance evaluation to drive operational excellence. By the conclusion of this chapter, readers will gain a comprehensive understanding of the role of predictive maintenance in enhancing asset management and operational efficiency, thereby empowering organizations to thrive in the dynamic landscape of industrial operations.

2. From laboratory to industrial settings

For finding trends in asset signal readings (see **Figure 1**) which indicate a potential failure in the system, the most convenient way would be to train a prediction model

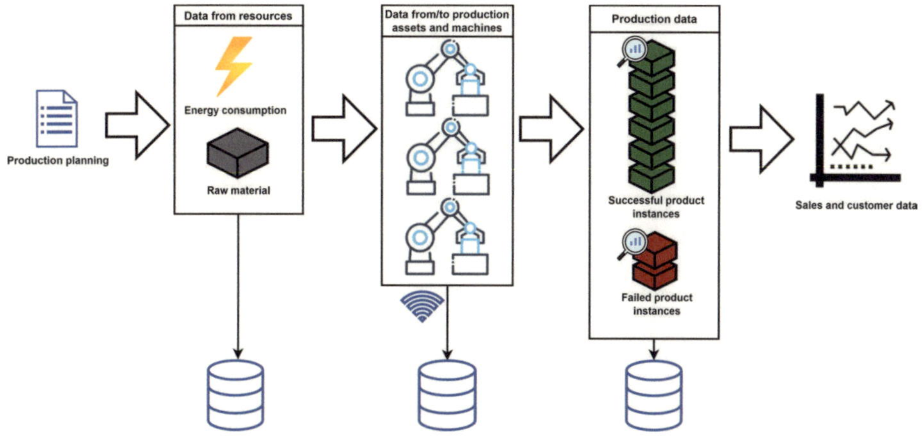

Figure 1.
Data acquisition from different assets in a smart factory.

with abundant samples of such sequences. For ensuring the generalization power of the prediction model for all the impending failures in the system, the gathered samples should cover different failure types. Once such data are acquired from the system, a classification or regression model can be trained to estimate and/or predict its status. Nevertheless, these samples are normally hard to attain in industry, as failures in such settings mean loss of productivity, reduced throughput, maintenance costs, and replanning costs. Thus, predictive maintenance solutions tend to be different in industry from the ones developed in perfect conditions and laboratories. In what follows, namely in Section 2.1, some of the reasons for lack of data and also annotated data are introduced.

Additionally, all decisions made in an industrial plant have to show their potential in having the return of investment as fast as possible. Therefore, for such use cases, it is vital to offer solutions that first of all do not cause substantial costs for sensor installation and asset verification (see Section 2.2) and can be also verified, given the operator's domain knowledge about the system (see Section 2.3).

Furthermore, given the constant changes in the production line, the deployed predictive maintenance solution must be aware of and robust towards these frequent changes impacting the read data. In Section 2.4, issues causing the aforementioned changes are discussed in detail.

In what follows, the briefly introduced concepts will be explained for a better comparison between predictive maintenance solutions developed for the industry and the ones developed in the controlled settings of a lab.

2.1 Lack of failure samples

In numerous industrial predictive maintenance settings, no instances of failures from the targeted asset are available, which could be due to the following:

1. The targeted asset is new and thus no annotated data have been previously gathered from it.

2. The industry owner did not store any historical data from the production line and/or did not annotate the data.

3. Gathering failure sample from the asset cannot be financially justified as the costs of developing the predictive maintenance solution would be higher than preventive maintenance.

4. Incidents leading to a failure in the asset can be dangerous for operators and/or users of the asset such as commercial planes.

In what follows, different strategies for dealing with lack of annotated for predictive maintenance are presented.

2.1.1 Predictive maintenance for new assets

There are several possible cases where a predictive maintenance solution has to be developed or adapted for a new asset. Occasionally, given the wear and tear during production, despite the normal usage of the asset, or the release of new series of the aforementioned asset, the target asset needs to be changed to increase productivity during production.

The former is easier to deal with as principally, the new replacement for the asset, has the same build and logic. In a simplified case, the similarity in the build and the logic of the asset must result in the same signal readings from the system as the physical characteristics of the asset have not changed and the behavior of the asset, controlled by the logic unit of the asset, is also as it previously was. Nonetheless, for more complex cases, no two instances of the same assets behave exactly the same, as assets after production go through calibration steps to ensure their required performance. Apart from the build and logic of the asset, there are numerous factors that potentially impact the gathered data from the asset, which are shown in **Figure 2**, but are not relevant for the current case and will be discussed further in Section 2.4.

On a separate note, it is also the case for many SMEs that they assign new tasks for different assets, as either they are not being used for production or priorities require dedicating more production assets for a specific customer. In this case, the build of the asset has not changed but its logic has changed, given the new production instructions. Regardless of the sources of changes in the read data from an asset, in case that annotated historical data from the previous asset are available, it is possible to adapt the available predictive maintenance solution to mitigate the changes in the read data from the new asset. In fact, this issue is a well-studied area of research in machine

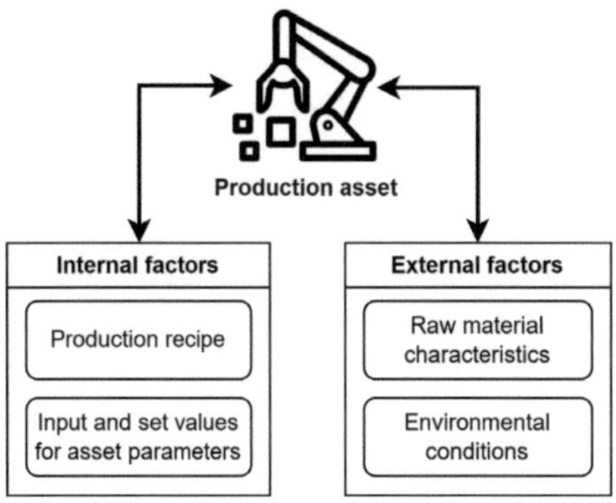

Figure 2.
Internal and external factors impacting the data gathered from a production asset.

learning called domain adaptation. In short, domain adaptation seeks to learn a model from a provided source of annotated data, which can later be generalized to a target domain by minimizing the difference between the source and target distributions (see **Figure 3**) or by relying on features that are source independent [1–4].

On the other hand, while developing a predictive maintenance solution for a new production line or a new asset, from which no former historical data are available, there are two possible ways to deal with lack of annotated data:

1. Using physical model of the system for data generation.

2. Using anomaly detection.

If the complexity of the targeted asset does not prevent creating a physical model, capable of recreating its behavior, the aforementioned model can be used to create annotated data [5, 6]. Nevertheless, one important note here is that, it must be feasible to create different scenarios in the simulation model which represent different failure types in the physical system. Otherwise, the data generated using the physical model can be used as a basis for the anomaly detection solution introduced later (**Figure 4**).

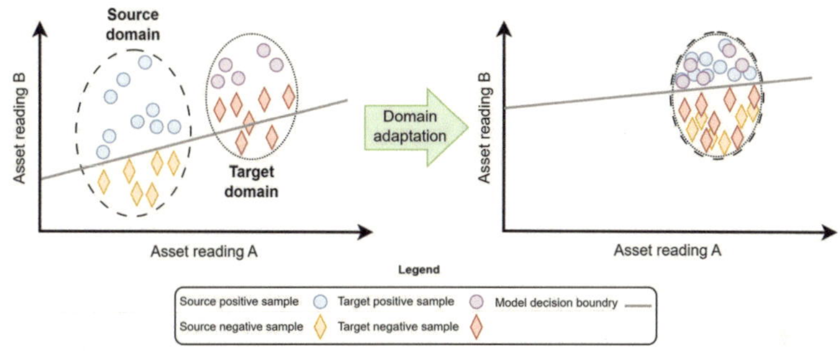

Figure 3.
Reducing distribution gap between the source and target domains by domain adaptation.

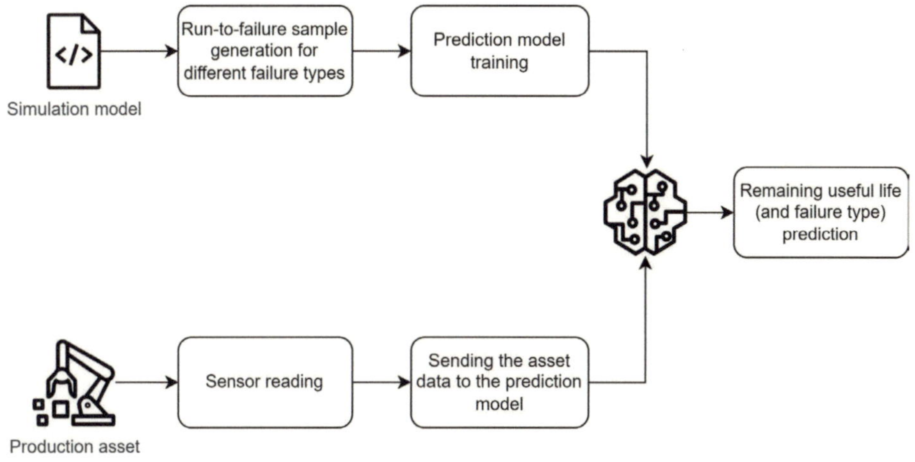

Figure 4.
Simulation model data generation for predictive maintenance.

Anomaly detection can be used as a basis for any new predictive maintenance solution, as a main building block of condition monitoring of the asset. Concisely, anomaly detection aims to flag parts of the signal readings from the asset that deviate from a previously provided pattern [7]. Moreover, anomaly detection as one of the most important use cases of unsupervised learning does not require annotated data and thus can significantly facilitate creating a predictive maintenance solution for a new asset or new production settings. One of the requirements for this method is a reference behavior from the system. As soon as this reference is available, different machine learning algorithms, such as autoencoder, one-class support vector machine, isolation forest, and different clustering algorithms, can be used to detect datapoints that are deviating from the provided reference behavior. Nonetheless, although this method may appear favorable, solutions relying on anomaly detection are mostly heavily dependent on their hyperparameters and metrics. Such hyperparameters help determine if a given datapoint is *different* or *far away* enough from the reference behavior, given an arbitrary *similarity* or *distance* measurement as the metric [8].

In addition, as data readings from industry are not perfect, usually it is not easy to find the perfect decision boundary separating the datapoints representing the normal working condition of the asset from the erroneous conditions and failures, given the signal readings. This exacerbates further when no samples of asset failure are available as there can be infinitely many decision boundaries that divide the available data space, given the datapoints representing the healthy status of the asset. Furthermore, decision boundary in different use cases has to weigh the importance of finding signal readings representing a failure against falsely flagging a normal working condition datapoint as erroneous (see **Figure 5**).

The aforementioned importance can be simply be translated into an objective function during the model training. Model improvement after deployment is also possible when more information about the importance of different classes of data is evident. Some values, which can help interpret the results of the prediction, are as follows [9]. Please note that the negative class represents scenarios where the target asset does not have any problem and is functioning as expected. The positive class, on the other hand, represents faulty asset states.

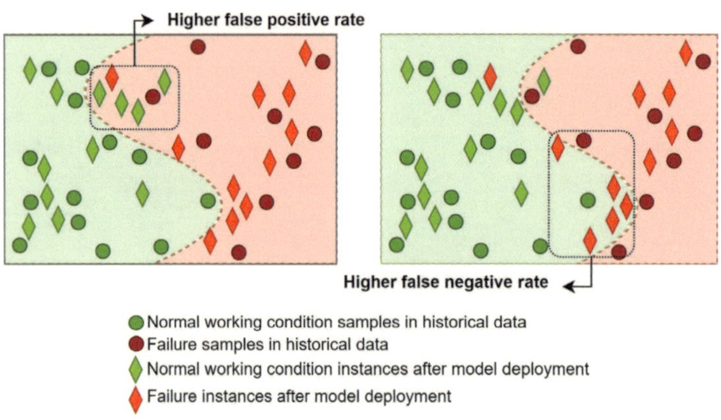

Figure 5.
Impact of decision boundary on false positive and false negative rates.

- True positive rate (TPR):

$$\frac{\#\, failure\ instances\ correctly\ classificed}{total \#\, failure\ samples}$$

- True negative rate (TNR):

$$\frac{\#\, normal\ instances\ classified\ correctly}{total \#\, normal\ samples}$$

- False positive rate (FPR): $1-TNR$

- False negative rate (FNR): $1-TPR$

- Accuracy:

$$\frac{\#\, correctly\ classified\ samples}{total \#\, samples}$$

- Balanced accuracy:

$$\frac{TPR+TNR}{2}$$

Given the highly unbalanced nature of data available in different predictive maintenance use cases (see Section 2.4), it is suggested to use accuracy measurement criteria such as balanced accuracy over a simple accuracy value to have a better overview on the performance of the deployed prediction model for the asset. In fact, relying on only one class of data in the available asset readings is not recommended as the ratio and importance of positive and negative classes can change in time, leading to numerous issues caused by biased sample selection and model training. By monitoring the overall performance of the system, *e.g.*, balanced accuracy, it is possible to adapt the hyperparameters of the solution and/or retrain the data-driven model, given the available data and the new insight for under/oversampling and readjusted sample weights. Interested readers are suggested to read the implementation concepts introduced by Continual learning for maintaining the performance of a data-driven solution, despite the possible data distribution shifts [10].

2.1.2 Lack of physical model of the system

Production assets frequently used in OEMs and SMEs are constantly growing more complex for increased efficiency and performance, making them harder to model. Therefore, in the recent studies addressing the lack of annotated data from an asset, the focus has shifted from gathering data from a simulation model for model training to anomaly detection. With the help of anomaly detection, it is possible to constantly compare the asset readings with a previously provided normal behavior and flag parts of the readings as anomaly which deviate from the expected trend [11]. The aforementioned deviation can also be translated into a health index for the given asset. In fact, the health index can be interpreted as a probability value which indicates how probable it is that the target asset in its current condition is defective. The calculated health indices can then be used to calculate the degradation of the asset in time. In essence, what follows are the steps recommended to implement an anomaly detection-based predictive maintenance solution (**Figure 6**):

1. Gathering asset readings representing normal working condition of an arbitrary asset.

2. Training an anomaly detection model, given the acquired normal working condition samples.

3. Deploying the anomaly detection model and calculating the deviation from the provided reference asset behavior.

4. Converting deviation values to health indices.

5. Predicting the future health index values for the asset for calculating its remaining useful life.

6. Planning maintenance.

Such an approach reduces the impact of the absolute value of the asset readings and puts more emphasis on the deviation from the expected values. Furthermore, in a rare case of a failure in the asset, the acquired datapoint can be used to update the thresholds and the different hyperparameters of the deployed model.

In the upcoming sections, more complex implementation and model evaluation topics for predictive maintenance solutions are introduced.

2.2 Efficient predictive maintenance: reducing sensors

There are several factors impacting the deployment costs of a predictive maintenance solution (see **Figure 7**). One of the main factors preventing SMEs or OEMs for integrating predictive maintenance in their ecosystem is related to the costs and effort associated with sensor installation. Especially, for companies that own fleets of production assets, installing additional sensors on each of these units can be time-consuming and expensive. The issues with sensor installation are further

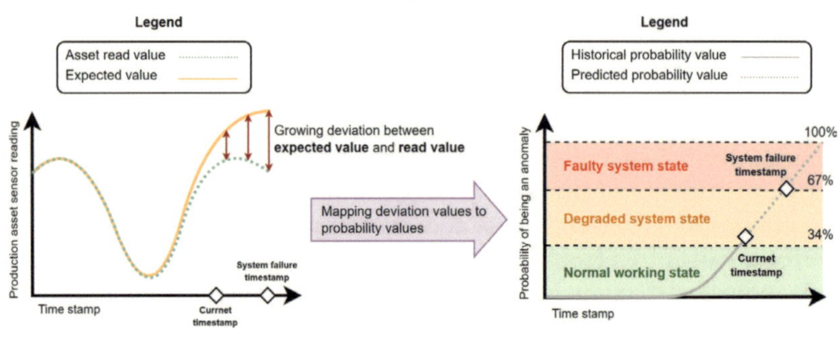

Figure 6.
Anomaly detection-based predictive maintenance.

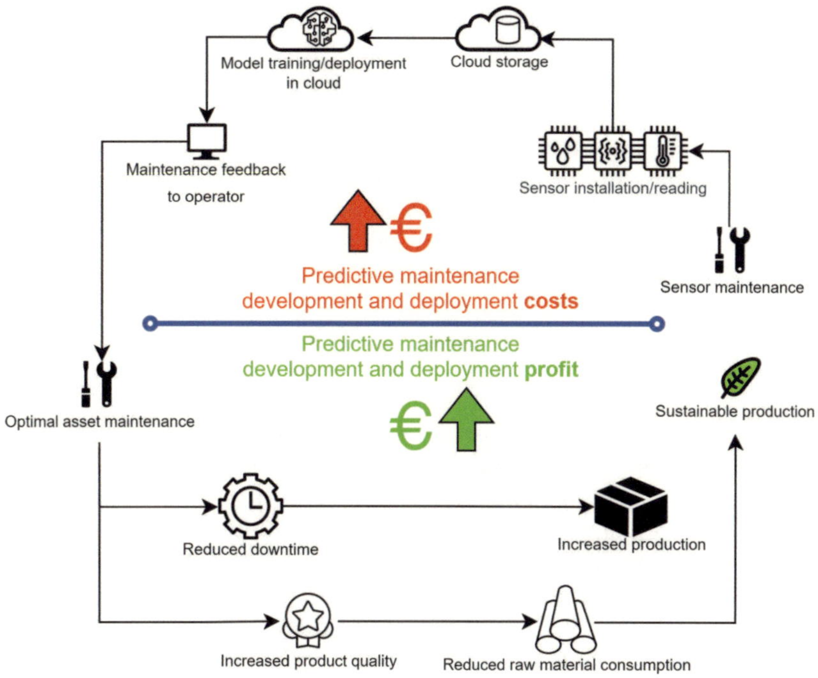

Figure 7.
Predictive maintenance costs and profit.

exasperating for pharmaceutical companies or producers of safety-critical system as the production assets have to be certified again and also reevaluated, given the adjustments made in the asset during sensor installation. This process can reduce the productivity particularly at the beginning of deploying the solution. Apart from the effort needed for sensor installation, as the information acquired from sensors is the very foundation of decisions made by the predictive maintenance solution, it is of utmost importance to ensure the quality of the data fed to it. Noisy, incomplete, and high latency in sensor reading are some of the common issues deteriorating the data quality and consequently the effectiveness of the developed predictive maintenance solution [12]. As a result, to prevent issues with data readings from the

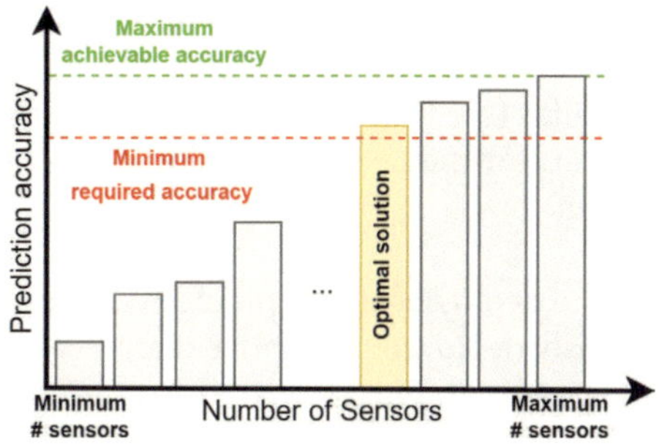

Figure 8.
Cost-effective predictive maintenance.

sensors, they need to be maintained as well, which could potentially lead to a never-ending loop of maintenance.

One effective way for reducing the number of sensors is by inspecting the impact of different sensors on the predictions made by the predictive maintenance solution. Thereafter, sensor readings can be chosen if their roll in the predictions can be physically justified, given the domain knowledge from the asset. In addition, such important readings must also have a significant contribution to the accuracy of the prediction model (see **Figure 8**), so that their selection can also be validated from a deployment cost point of view. This approach is thoroughly examined in the next subsection.

2.3 Interpretable predictive maintenance: verifying prediction models with domain knowledge

An important characteristic of data-driven predictive maintenance solutions ensuring their acceptance in industry is their interpretability. Model interpretability can help verify the trained prediction model and have an overview of how it uses different sources of information from the asset [13]. In fact, by using different techniques introduced in interpretable AI, it is possible to infer the role of different asset readings on the decisions made by the prediction model and in case of a wrong prediction, the detected error can be elucidated, given this information. Provided that the deployed predictive maintenance solution complies with the requirements of trustworthy AI [14], it is also possible to examine which datapoints were used to train the model that led to the wrong prediction of the model and prevent similar incidents in the future. Under the circumstances that some domain knowledge is available from the maintenance crew of the SME or OEM, it is also possible to visualize the feature importance of the prediction model and have the experts verify the behavior of the model. In addition, by inspecting the importance of different asset readings, it is feasible to inspect if there are any specific readings that dictate the output of the prediction model and whether or not these sensor readings can be relied on under different working conditions. For ensuring the acceptable performance of the predictive maintenance solution, it is recommended to have a prediction model that generates its output, given a wider range of sensor readings with no dominant peak in importance values (see **Figure 9**).

2.4 Maintaining predictive maintenance models in industry

One of the main goals of Industry 4.0/5.0 is to adapt different assets of a production line to meet the requirements of the end user. This adaptability in production often leads to changes in the read asset data [15, 16]. Therefore, for a successful and impactful deployment of a predictive maintenance solution it is crucial to develop measures for maintaining the prediction model [17] and protecting it from different types of data distribution shifts (see **Figure 10**). Some of the most evident distribution shifts or potential reasons for them in different predictive maintenance implementations are the following:

1. Imbalanced data: prior to deploying a new predictive maintenance solution, the availability of a previously deployed preventive maintenance solution can impact the distribution of datapoints attained from the asset from different working conditions. If the deployed preventive maintenance solution had a low performance or was nonexistent, numerous instances of system failure could have been gathered. On the other hand, if the allocated preventive maintenance solution hindered

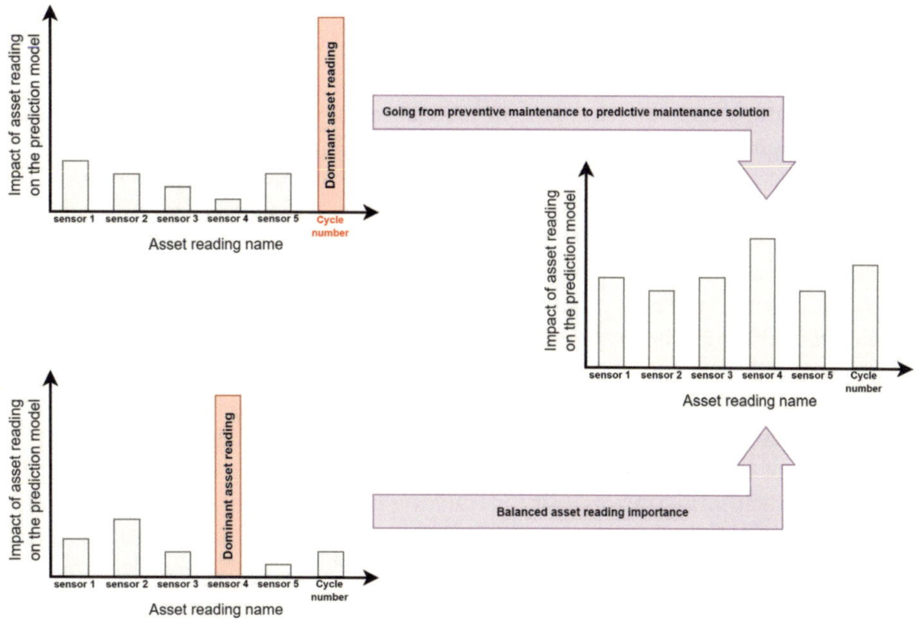

Figure 9.
Role of model interpretability on the evaluation of a predictive maintenance solution.

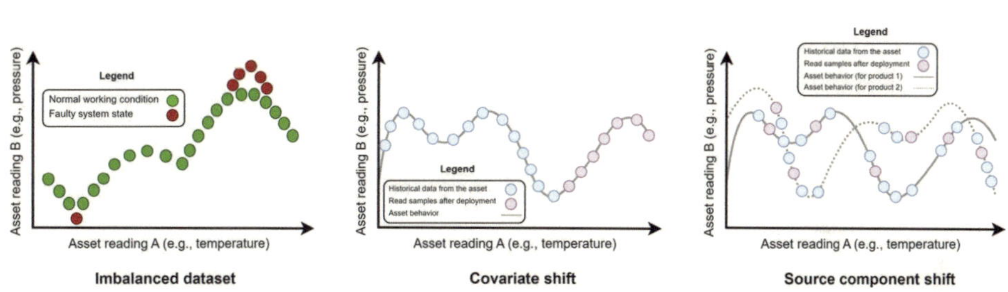

Figure 10.
Data imbalance and distribution shifts in predictive maintenance.

gathering samples that represent different failure types in the targeted asset, then such failure samples would be scarce. The ratio of datapoints representing normal working condition and erroneous states of the system will ultimately change when an effective predictive maintenance solution is deployed. The unbalance in the data should always be accounted for in different parts of the data pipeline and model training by either upweighting some of the gathered samples from the production asset or by under/oversampling the instances of some working conditions. In fact, imbalanced data reflect sample selection bias with a known bias value that is determined by the label of the class gathered from the asset [18].

2. Covariate shift: given the possible changes in the working condition and also the external factors from the production shop floor impacting an asset, these changes could potentially have not been reflected in the data used to train a predictive maintenance solution. Therefore, the model is not qualified to make predictions for such new production settings. If the aforementioned changes are not tracked

for evaluating the predictions of the trained model, they could be misleading as the trained model is working under different assumptions [19].

3. Source component shift: in many industrial settings, different production assets are reassigned for different tasks than they were initially used for. The new production recipe in this case will result in data with different characteristics. For such occurrences, the previously trained model is again obsolete and needs to be adapted to the new production circumstances. As soon as different models for different production recipes are available, the next step would be to first identify which task the read asset data are indicating and then use the correct prediction model to evaluate the production asset status [20].

As it can be seen, there are numerous internal and external factors impacting the gathered data from an asset. This constant change in the data influences the performance of the data-driven predictive maintenance solution. Thus, it is inevitable to have an administration unit for tracking the changes in the deployed models and also to log the changes in the data for future references. One potential solution for fulfilling this task is the AAS [21], which is introduced in the next section.

3. AAS as a potential performance booster for predictive maintenance

The landscape of predictive maintenance is continuously evolving, driven by advancements in technology and the increasing complexity of industrial operations. As organizations strive to optimize asset management and minimize downtime, the concept of the AAS emerges as a potential game-changer in enhancing the performance of predictive maintenance solutions. In this section, we delve into the role of AAS as a facilitator of interoperability, data exchange, and performance enhancement in the context of predictive maintenance.

3.1 Understanding the Asset Administration Shell

Asset Administration Shell (AAS) represents a standardized framework within the context of the RAMI4.0. At its core, the AAS provides a digital representation of physical assets and their associated data, enabling seamless integration and communication across heterogeneous systems. By encapsulating comprehensive information about an asset, including its structure, behavior, and lifecycle data, the AAS fosters interoperability and transparency within industrial environments. A detailed description of the AAS information model can be found in the specification (see **Figure 11**) [22]. The AAS encapsulates comprehensive information about each asset, including its structure, behavior, and lifecycle data. This information is structured according to a predefined information model that defines the attributes and relationships necessary to describe an asset comprehensively. By adhering to this standardized format, the AAS ensures interoperability and transparency, enabling stakeholders to exchange data and information effectively. One of the key aspects of the AAS is its ability to establish digital twins and virtual representations of physical assets. These digital twins serve as virtual counterparts to their physical counterparts, providing real-time insights into asset health, performance trends, and maintenance requirements. By leveraging digital twins, organizations can monitor

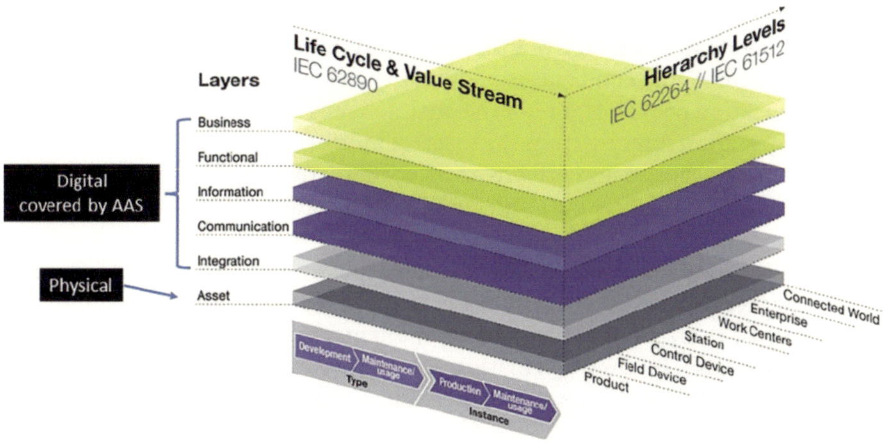

Figure 11.
Three-dimensional (3D) representation of the reference RAMI4.0 based on [23].

assets remotely, identify anomalies or potential issues, and optimize maintenance schedules proactively.

3.2 Leveraging AAS for predictive maintenance

Integrating the principles of AAS into predictive maintenance initiatives holds immense potential for enhancing the efficiency and effectiveness of asset management strategies. Through the establishment of digital twins and virtual representations of physical assets, organizations can gain insights into asset health, performance trends, and maintenance requirements in real-time. By adopting the AAS framework, organizations can achieve several benefits in the context of predictive maintenance:

- *Data integration and aggregation*: the AAS enables the aggregation of disparate data sources, including sensor data, maintenance records, and operational parameters, into a unified digital representation. These aggregated data serve as the foundation for predictive analytics and machine learning algorithms, empowering organizations to extract actionable insights and make informed decisions regarding asset maintenance and optimization.

- *Enhanced transparency*: the AAS enables a high level of transparency regarding asset status, performance metrics, and maintenance requirements. By encapsulating all relevant information within the asset's administration shell, stakeholders gain comprehensive visibility into the asset's condition and operational history. This transparency facilitates informed decision-making regarding maintenance scheduling, resource allocation, and asset optimization strategies.

- *Improved data accessibility*: the AAS facilitates the centralized storage and management of asset-related data, ensuring easy accessibility for relevant stakeholders across the organization. With data stored in a standardized format within the asset's administration shell, authorized personnel can retrieve and analyze critical information efficiently. This accessibility streamlines data-driven decision-making processes, enabling timely interventions and proactive maintenance activities.

- *Integration with predictive maintenance systems*: the AAS framework can seamlessly integrate with predictive maintenance systems, providing them with access to real-time asset data and status updates. By leveraging AAS interfaces and communication protocols, predictive maintenance solutions can retrieve pertinent information directly from the asset's administration shell. This integration enhances the accuracy and effectiveness of predictive maintenance models by incorporating up-to-date data and contextual insights.

- *Interoperability and interconnectivity*: by adhering to standardized AAS formats and interfaces, predictive maintenance solutions can seamlessly communicate with diverse industrial systems and components. This interoperability facilitates the exchange of data and information across the entire value chain, enabling enhanced collaboration between stakeholders and ensuring holistic asset management strategies.

- *Enabling data-driven insights*: with the comprehensive data stored within the asset's administration shell, organizations can harness advanced analytics and machine learning algorithms to derive actionable insights. By analyzing historical performance data, identifying patterns, and predicting future maintenance needs, organizations can optimize asset utilization, minimize downtime, and enhance overall operational efficiency. The structured nature of data within AAS facilitates sophisticated analytics, enabling organizations to unlock valuable insights for predictive maintenance optimization.

- *Ensuring security and integrity*: the AAS incorporates robust security mechanisms to safeguard asset data and prevent unauthorized access or tampering. By implementing encryption, authentication, and access control measures, organizations can mitigate cybersecurity risks and ensure the integrity of asset-related information. This security framework is essential for maintaining trust and reliability in predictive maintenance systems, particularly in industrial environments where data confidentiality and integrity are paramount.

3.3 Case studies and success stories

Real-world case studies and success stories demonstrate the tangible benefits of leveraging AAS for predictive maintenance across various industries. These case studies highlight the implementation challenges, best practices, and measurable outcomes achieved through the integration of AAS into predictive maintenance strategies. Examples include:

- *Automotive industry*: Volkswagen Sachsen GmbH implements AAS-based solutions to optimize production line efficiency, reduce unplanned downtime, and improve product quality. By leveraging digital twins and real-time data analytics, the automotive manufacturer achieves significant cost savings and operational enhancements across its manufacturing facilities [24].

- *Manufacturing sector*: Hitachi, a multinational semiconductor manufacturer, uses the AAS as a component of its predictive maintenance framework, enabling proactive asset management and predictive analytics. By standardizing asset data formats, optimizing data exchange processes, and implementing algorithms for predictive maintenance, the company achieves new levels of operational efficiency and asset reliability, resulting in higher and consistent product quality [25].

3.4 Future directions and emerging trends

Looking ahead, the integration of AAS into predictive maintenance is poised to drive further innovation and transformation in industrial asset management. Emerging trends, such as edge computing, artificial intelligence, and digital twins, promise to revolutionize predictive maintenance capabilities, offering organizations new opportunities for optimization and competitive advantage.

Edge computing, for instance, brings computational power closer to the data source, enabling real-time analysis and decision-making at the network's edge. By deploying edge computing solutions in conjunction with AAS-enabled predictive maintenance, organizations can enhance responsiveness and reduce latency, thereby improving overall operational efficiency.

Artificial intelligence (AI) plays a pivotal role in augmenting predictive maintenance capabilities by enabling advanced analytics, anomaly detection, and predictive modeling. By leveraging AI algorithms, organizations can extract valuable insights from large volumes of data generated by AAS-enabled systems, facilitating proactive maintenance interventions and optimizing asset performance.

Digital twins represent another transformative trend in predictive maintenance, offering virtual replicas of physical assets that mirror their real-world behavior and characteristics. By creating digital twins within the AAS framework, organizations can simulate asset performance, conduct scenario analysis, and optimize maintenance strategies in a risk-free virtual environment.

As organizations continue to embrace digitalization and Industry 4.0 initiatives, the integration of AAS into predictive maintenance strategies will play a pivotal role in shaping the future of asset management and maintenance practices. By keeping up with these trends and embracing technological advancements, organizations can fully exploit the capabilities of predictive maintenance solutions enabled by AAS, driving continuous improvement and innovation in industrial operations.

4. Predictive maintenance in OEM vs. SME: differences and pivoting points

Currently, the main driving forces for implementing predictive maintenance for different industrial settings are reducing the downtime, avoiding micro-stops, and increasing production. However, one of the more important side benefits of predictive maintenance that is normally neglected in the related literature is the impact of production asset state on the produced product quality [26]. As an example, the wear and tear in pneumatic actuators moving the tool center point of a production asset influences highly the precision of the production asset, which as a result deteriorates the final product. For both OEMs and SMEs, it is crucial to reduce the scrap rate during production for minimizing the raw material consumption and decreasing the production cost. Apart from production cost reduction, by maintaining the production assets in an acceptable condition, it is possible to ensure certain final product quality and prevent the need for providing replacements for instances that did not meet the customer's needs. Furthermore, avoiding high scrap rate also reduces the energy consumption during production which boosts sustainable production.

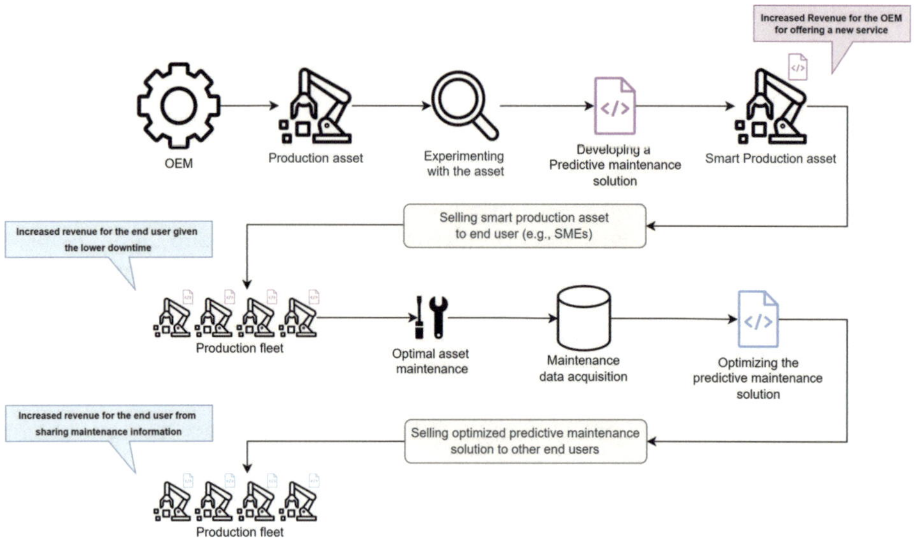

Figure 12.
Smart asset and maintenance solution sharing for increasing revenue.

Apart from boosting performance and sustainability in production, OEMs can use predictive maintenance as an important source of income in their business model. By offering smart production assets with functioning predictive maintenance solutions, the end user will indirectly benefit from the OEM's expertise during production for reaching their goal with comparably less effort and resource use. In addition, as soon as the end user has gathered enough data during production, they can adapt the provided generic predictive maintenance solution from the OEM to better suit their needs and also analyze the impact of the production asset on the quality of the produced product. Once the aforementioned datasets are produced, the end user can also integrate selling data to other end users with the similar production assets to their business model (see **Figure 12**). More details regarding privacy and data protection are provided in the next section.

5. Privacy and data protection in predictive maintenance aimed towards the end customer

With the continuous advancement of data-driven solutions in different areas of research, numerous key figures in industry have grasped the significance of proper digitalization and data acquisition for their assets. The importance, effort, and costs associated with industrial data have boosted their value up to a point that some business models in industry suggest sharing and selling industrial data as a crucial source of income, especially for OEMs. Nonetheless, one major ethical question that is raised here, is

> *Once a customer buys a production asset, e.g., a CNC (computer numerical control) machine, and uses this asset for their own production, who is the owner of the data attained from the initial production asset?*

To protect the end user of a production asset and also to enable OEMs to benefit from their efforts in producing smart production assets, there are multiple ways to

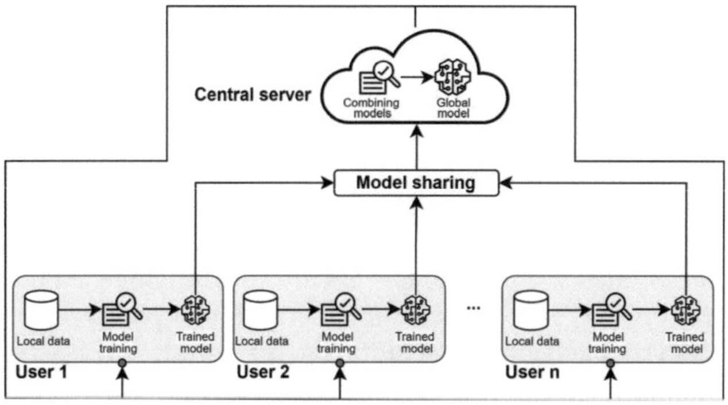

Figure 13.
Preventing direct data sharing via federated learning.

indirectly share data between different users without compromising privacy. Possible solutions include but are not limited to data encryption [27] and federated learning [28]. By encrypting the data, no raw data will be shared among different users protecting different companies from their production plans to be revealed. However, the data-driven models used in these scenarios need to be capable of using the encrypted data from different sources in a way that boosts their performance.

Another method for sharing production asset, without directly sharing the data from different users, is to employ federated learning. By using this method, the trained models from different users are shared and retrained using the data from other customers (see **Figure 13**). In case the model accuracy increases, the newly trained model will be adapted by the initial user. This solution does not require actual sharing of the data and only the trained models will be passed around. Nevertheless, it is possible to attain the initially used dataset to train a model from many solutions using deep neural networks. Therefore, in such scenarios, measures needed to protect the data from being extracted from the trained model need to be taken to ensure maximum data privacy for the end user.

6. Conclusion

In this chapter, different internal and external factors impacting various aspects of planning, implementing, and deploying effective predictive maintenance solutions were discussed. It was shown how the availability of data, data quality, and data annotation can steer the design process. Furthermore, important concepts in trustworthy AI were introduced, which can help maintain and boost the performance of the deployed predictive maintenance solutions. It was also explained in detail how model interpretability can help reduce the required number of sensors and thus lower the deployment costs, motivating different customers to equip their production assets with sensors and monitor their efficiency. In addition, AAS as a promising standardization for different Industry 4.0/5.0 use cases in Europe was introduced. It was discussed how AAS can help track asset data in different industrial settings, and how the unified digital representation can then be used in the data pipeline and also model training. The standardized data attained from different assets can ultimately enhance the performance of the deployed data-driven solutions. Afterwards, it was shown how predictive maintenance can impact the business model of OEMs and

SMEs, as they can earn additional money just by providing the foundation needed for gathering data from the production assets and later share the data with other users as a service. Lastly, possible issues with direct data sharing were pointed out and federated learning as a feasible solution for improving predictive maintenance models without data sharing was introduced. The reader of this chapter must now have a broad overview on the most demanding and complicated issues for developing and deploying predictive maintenance solution in industry. Depending on their need, the reader can further pursue different topics that are currently needed in industry, such as model lifecycle support, edge device implementation of prediction models, and local and cloud data preparation and structuring for production assets.

Acknowledgements

The authors of this book chapter would like to express their deep sense of gratitude to SMC Schweiz AG, especially Mr. Alessandro Grizzetti, for their unceasing assistance in promoting Competence Centre for Automation and Digitalisation among different industrial partners in Europe.

Conflict of interest

The authors declare no conflict of interest.

Abbreviations

OEM	original equipment manufacturer
SME	small and medium-sized enterprise
IoT	Internet of Things
AAS	Asset Administration Shell
RAMI4.0	Reference Architecture Model Industrie 4.0
TPR	true positive rate
TNR	true negative rate
FPR	false positive rate
FNR	false negative rate
AI	artificial intelligence

Author details

Kiavash Fathi[1,2] and Hans Wernher van de Venn[1*]

1 Zurich University of Applied Sciences, Winterthur, Switzerland

2 Information and Automation Systems for Process and Material Technology, RWTH Aachen University, Aachen, Germany

*Address all correspondence to: vhns@zhaw.ch

References

[1] Farahani A, Voghoei S, Rasheed K, Arabnia HR. A brief review of domain adaptation. Advances in Data Science and Information Engineering: Proceedings from ICDATA 2020 and IKE 2020. 2021:877-894

[2] Zhou K, Liu Z, Qiao Y, Xiang T, Loy CC. Domain generalization: A survey. IEEE Transactions on Pattern Analysis and Machine Intelligence. 2022;**45**(4):4396-4415

[3] Nejjar I, Geissmann F, Zhao M, Taal C, Fink O. Domain adaptation via alignment of operation profile for remaining useful lifetime prediction. Reliability Engineering & System Safety. 2024;**242**:109718

[4] Rahat M, Mashhadi PS, Nowaczyk S, Rognvaldsson T, Taheri A, Abbasi A. Domain adaptation in predicting turbocharger failures using vehicle's sensor measurements. Phm Society European Conference. 2022;7:432-439

[5] Didona D, Quaglia F, Romano P, Torre E. Enhancing performance prediction robustness by combining analytical modeling and machine learning. In: Proceedings of the 6th ACM/SPEC International Conference on Performance Engineering. 2015. pp. 145-156. DOI: 10.1145/2668930.2688047

[6] Lei Y, Li N, Gontarz S, Lin J, Radkowski S, Dybala J. A model-based method for remaining useful life prediction of machinery. IEEE Transactions on Reliability. 2016;**65**(3):1314-1326

[7] Chandola V, Banerjee A, Kumar V. Anomaly detection: A survey. ACM Computing Surveys (CSUR). 2009;**41**(3):1-58

[8] Fathi K, van de Venn HW, Honegger M. Predictive maintenance: An autoencoder anomaly-based approach for a 3 dof delta robot. Sensors. 2021;**21**(21):6979

[9] Monaghan TF, Rahman SN, Agudelo CW, Wein AJ, Lazar JM, Everaert K, et al. Foundational statistical principles in medical research: Sensitivity, specificity, positive predictive value, and negative predictive value. Medicina. 2021;**57**(5):503

[10] Hurtado J, Salvati D, Semola R, Bosio M, Lomonaco V. Continual learning for predictive maintenance: Overview and challenges. Intelligent Systems with Applications. 2023;**19**:200251

[11] Carrasco J, López D, Aguilera-Martos I, García-Gil D, Markova I, Garcia-Barzana M, et al. Anomaly detection in predictive maintenance: A new evaluation framework for temporal unsupervised anomaly detection algorithms. Neurocomputing. 2021;**462**:440-452

[12] Renggli C, Rimanic L, Gürel NM, Karlaš B, Wu W, Zhang C. "A data quality-driven view of mlops". 2021. arXiv preprint arXiv:2102.07750

[13] Steurtewagen B, Van den Poel D. Adding interpretability to predictive maintenance by machine learning on sensor data. Computers & Chemical Engineering. 2021;**152**:107381

[14] Li B, Qi P, Liu B, Di S, Liu J, Pei J, et al. Trustworthy ai: From principles to practices. ACM Computing Surveys. 2023;**55**(9):1-46

[15] Sculley D, Holt G, Golovin D, Davydov E, Phillips T, Ebner D, et al. Hidden technical debt in machine learning systems. Advances in Neural Information Processing Systems.

2015;**28**. Available from: https://proceedings.neurips.cc/paper_files/paper/2015/hash/86df7dcfd896fcaf2674f757a2463eba-Abstract.html

[16] Paleyes A, Urma R-G, Lawrence ND. Challenges in deploying machine learning: A survey of case studies. ACM Computing Surveys. 2022;**55**(6):1-29. Available from: https://proceedings.neurips.cc/paper_files/paper/2015/hash/86df7dcfd896fcaf2674f757a2463eba-Abstract.html

[17] Kreuzberger D, Kühl N, Hirschl S. Machine learning operations (mlops): Overview, definition, and architecture. IEEE Access. 2023. pp. 31866-31879

[18] Quiñonero-Candela J, Sugiyama M, Schwaighofer A, Lawrence ND. Dataset Shift in Machine Learning. Mit Press; 2008. Available from: https://mitpress.mit.edu/9780262545877/dataset-shift-in-machine-learning/

[19] Huyen C. Designing Machine Learning Systems. O'Reilly Media, Inc; 2022. Available from: https://www.oreilly.com/library/view/designing-machine-learning/9781098107956/

[20] Fathi K, Sadurski M, Kleinert T, van de Venn HW. Source component shift detection & classification for improved remaining useful life estimation in alarm-based predictive maintenance. In: 2023 23rd International Conference on Control, Automation and Systems (ICCAS), Yeosu, Korea. IEEE; 2023. pp. 975-980

[21] Cavalieri S, Salafia MG. A model for predictive maintenance based on asset administration shell. Sensors. 2020;**20**(21):6028

[22] Bader S, Barnstedt E, Bedenbender H, Berres B, Billmann M, Ristin M. "Details of the Asset Administration Shell-Part 1: The Exchange of Information Between Partners in the Value Chain of Industrie 4.0 (Version 3.0 rc02)". Berlin, Germany: Federal Ministry for Economic Affairs and Climate Action (BMWK); 2022

[23] Spec D. 91345: Referenzarchitekturmodell industrie 4.0 (rami4. 0). Deutsches Institut für Normung (DIN) eV. 2016

[24] Case IU. "A Uniform Information Model for Manufacturing". 2024. [Online]. Available from:: https://iotusecase.com/en/use-cases/a-uniform-information-model-for-manufacturing/

[25] Fathi K, Stramaglia M, Ristin M, Sadurski M, Kleinert K, Schönfelder R, et al. 12th IFAC Symposium on Fault Detection, Supervision and Safety for Technical Processes: Safeprocess 2024: 4-7 June 2024: Proceedings, Ferrara, Italy. Elsevier Science Direct - IFAC PapersOnline; 2024

[26] Acebo AG, Ventura-Traveset A, Uriarte S d l M. Remote condition monitoring and control for maintenance activities in coordinate measurement machine: A data-driven approach for the metrology engineering industry. Predictive Maintenance in Smart Factories: Architectures, Methodologies, and Use-cases. 2021:121-147. Available from: https://link.springer.com/chapter/10.1007/978-981-16-2940-2_6#citeas

[27] Podschwadt R, Takabi D, Hu P, Rafiei MH, Cai Z. A survey of deep learning architectures for privacy-preserving machine learning with fully homomorphic encryption. IEEE Access. 2022;**10**:117477-117500. Available from: https://link.springer.com/chapter/10.1007/978-981-16-2940-2_6#citeas

[28] Zhang C, Xie Y, Bai H, Yu B, Li W, Gao Y. A survey on federated learning. Knowledge-Based Systems. 2021;**216**:106775

Chapter 6

Contribution of Artificial Intelligence to Industrial Maintenance in the Field of Mechanics

Mohamed El Khaili, Mohamed Rafik, Redouane Fila and Abdelmajid Farid

Abstract

The global industry is in continuous technological evolution, which aims for reliability, efficiency, availability, and safety while reducing maintenance costs. Modern maintenance follows change, which can no longer be limited to being corrective or preventive, but must be proactive involving the continuous monitoring and verification of the root causes of failure; it must also be predictive which makes it possible to anticipate breakdowns and increase equipment usage time based on the Prognosis and Health Management (PHM), which transforms raw data into indicators and makes it possible to define the Residual Life (RUL) and its extrapolation as a decision-making tool. Our chapter consists of presenting the contribution of AI to industrial maintenance in the field of mechanics. It focuses on industrial maintenance through its concepts, technologies, and methods used. So, the presentation of artificial intelligence and its algorithms applied toward maintenance 4.0 are to show the contribution of AI to maintenance.

Keywords: predictive maintenance, Residual Life (RUL), prognosis as a service, Prognostics and Health Management (PHM), cloud computing, artificial intelligence (AI), performance measurements, Internet of Things (IoT), maintenance 4.0

1. Introduction

In a market that is continuously growing, and to remain competitive, industries are orienting their management policies toward increasing productivity at a lower cost while optimizing resources and setting availability, reliability, maintainability, and safety as objectives for the proper functioning of their production systems.

These objectives can be achieved through the implementation of an adequate and relevant maintenance strategy. This interest is fueled by the fact that an unplanned shutdown can have important and significant economic consequences for the company. In this context, and in a spirit of anticipation, companies are transforming themselves to extract value from traditional industrial sectors and make them more efficient than ever by providing data-rich digital services thanks to technological evolution as well as the development of artificial intelligence and automated production systems. By leveraging ultra-low-cost connectivity, there is rapid explosion of sensors, powerful analytical tools, as well as data storage capacity. So, it becomes necessary and important to involve these new technologies in industrial maintenance and to integrate artificial intelligence (AI), which has become crucial to ensure the proper functioning of industrial systems, and therefore, to achieve operational and financial efficiency.

Using AI, management can always have accurate information about the operating status of any machine or equipment. Thus, AI in predictive maintenance helps businesses save money and resources. This is done by adapting maintenance routines to the needs of each piece of equipment, rather than forcing them to follow a rigid schedule.

Artificial intelligence is one of these structural innovations that has an impact in many areas. If its roots go back to the 1950s, recent technological developments, including machine learning and deep learning, open new possibilities for using artificial intelligence in industrial maintenance. Investing in maintenance helps avoid repair costs that also lead to production downtime. Having an intelligent solution to determine the precise timing of each technical maintenance intervention on industrial assets constitutes a key competitive advantage by significantly reducing the cost of a critical and urgent repair, as well as avoiding possible interruptions of service necessary for the latter.

To make the transition to 4.0 maintenance, it is important to proceed gradually according to your starting level, your internal resources, and your appetite for digital technology. Indeed, each technological development involves profound changes that affect the organization, the teams, and the technical infrastructure. It is therefore necessary to audit and map internal processes as well as data flows.

The outline of this chapter is divided into three parts. In the first part, we will focus on industrial maintenance through its concepts, technologies, and methods used. The second part will be devoted to the presentation of AI and its algorithms applied toward maintenance 4.0. In the third part, we will study the contribution of AI to maintenance based on research into the history and statistics as well as trends in AI.

2. Industrial maintenance in the field of mechanics

2.1 Definition of industrial maintenance

According to the Association Française de NORmalisation (AFNOR) NF-X 60,000 [1] standard, industrial maintenance is defined as follows: *"It is all activities intended to maintain or restore an asset in a specified state or in given operational safety conditions, to accomplish a required function."*

Maintaining means preventing machine malfunctions, and restoring to correct following a machine malfunction. A specified state is what is defined by quantifiable characteristics and objectives. The industrial maintenance function revolves around different actions: management, diagnosis, troubleshooting, repair, verification, control, etc.

2.1.1 Maintenance concepts

The different maintenance concepts can be classified into three main categories: corrective maintenance, preventive maintenance, and predictive maintenance (**Figure 1**).

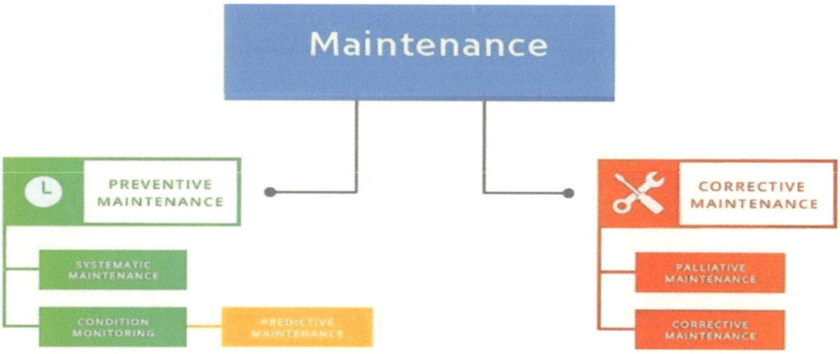

Figure 1.
Different forms of maintenance [1].

2.1.1.1 Corrective maintenance

The standard [1] defines corrective maintenance as: *"Maintenance carried out after detection of a breakdown and intended to restore an asset to a state in which it can perform a required function."*

Corrective maintenance (Reactive) is carried out after the occurrence of a fault in the system. It is generally adopted for equipment for which:

- the consequences of the breakdown are not critical,
- the repair is easy and does not require much time,
- and investment costs are low.

The concept of corrective maintenance aims to reset the system to its normal operating state after the occurrence of its failure.

2.1.1.2 Preventative maintenance

Preventive maintenance aims to reduce the risk of a failure occurring. Standard [1] defines it as follows: "Maintenance carried out at predetermined intervals or according to prescribed criteria and intended to reduce the probability of failure or degradation of the functioning of an asset." It is prepared and scheduled before the probable date of occurrence of a failure.

2.1.1.2.1 Systematic preventive maintenance

It is maintenance carried out according to a schedule established according to time or the number of units of use [1]. The frequency of replacements is determined using two methods: the first is block type, and the second is age type. The age-type replacement policy suggests replacing equipment at failure or after T units of uptime. The block-type policy suggests replacing the equipment after a predetermined period T, 2T, etc., regardless of the age and condition of the component. Systematic preventive maintenance can lead to over-maintenance, that is, to an excess of unnecessary interventions, and therefore to financial waste for the company. To compensate for this, other forms of preventive maintenance, based on monitoring the actual condition of equipment, have appeared, such as conditional maintenance [2].

2.1.1.2.2 Conditional preventive maintenance

According to standard [1], it is defined as: "Preventive maintenance based on monitoring of the operation of the asset and/or the significant parameters of this operation integrating the resulting actions."

2.1.1.3 Predictive maintenance

It aims to compensate for the lack of knowledge of condition-based maintenance. It is defined according to standard [1] as: "Conditional maintenance carried out following the predictions extrapolated from the analysis and evaluation of significant parameters of the degradation of the asset."

It also aims to offset the costs of corrective maintenance by minimizing the downtime of systems and, above all, by being able to plan these downtimes. This anticipatory method, therefore, makes it possible to ensure better continuity of service and thus reduces operating costs in the long term [3, 4].

Figure 2 shows the process of a predictive action.

It considers the current conditions of the equipment and attempts to predict the evolution over time of the condition of the property. The expected benefits are indeed numerous:

- Reduction in the number of breakdowns,

- Reliability of production,

Figure 2.
Process of a predictive action.

- Reduction of equipment downtime (costly),
- Increased business performance.

Predictive maintenance is based on continuous monitoring of system evolution to prevent shutdown before it happens [3, 4]. Predictive maintenance techniques help determine the condition of equipment in service to predict when maintenance should be performed. This approach saves money compared to routine or time-based preventative maintenance because tasks are performed only when warranted. The primary value of predicted maintenance is to enable convenient planning of corrective maintenance and avoid unforeseen equipment failures. The key is "the right information at the right time" (**Table 1**).

2.1.2 Evolution of maintenance

The effectiveness of the maintenance of industrial systems is a major economic issue for their commercial operation because the longer the maintenance phase, the more costly it is and generates unavailability of the system and a lack of productivity. The main difficulties and sources of inefficiency lie in the choice of type of maintenance (see **Figure 3**). The decision for a maintenance action is very complex and must be based on monitoring and intelligent analysis of the state of the system [1, 3].

2.2 Industrial maintenance technologies

Many companies supported by technological capabilities are improving their businesses. We see this in the various equipment maintenance processes, which are increasingly assisted by IT tools, cloud computing, Big Data, and the Internet of Things (IoT). Recently, IT and OT networks have significantly transformed industrial processes. Information technology (IT) and operational technologies (OT) are gradually transforming industrial organizations into digital businesses based on reliable data exchange.

One of the biggest influences on trends is Industry 4.0. Even before the pandemic, Industry 4.0 technologies were transforming the way manufacturers work. Today, the effectiveness of leaders in using technology is essential [5–8]. With the advent of Industry 4.0 [9, 10] in manufacturing, businesses can leverage new technologies to monitor and better understand their operations in real time, transforming a typical manufacturing facility into a smart factory.

Standard maintenance	**Maintenance**				
	Fix maintenance		**Preventative maintenance**		
	Palliative	**Curative**	**Systematic**	**Conditional**	**Predictive**
Trigger event	Failure	Failure	Date/deadline	Limit or threshold crossing	Drifts, trends
Action service	Repair	Repair	Systematic replacement	Conditional replacement	Targeted intervention

Table 1.
Type of maintenance according to triggering event.

Figure 3.
Evolution of maintenance.

In the context of predictive maintenance [11–13], AI makes it possible to analyze data, which is already used by other users or installers to anticipate certain operations. However, these risks resulting in a change in intervention processes among field operators.

Figure 4 shows the benefits of predictive maintenance.

2.3 Industrial maintenance methods

Today, technology plays an important role in many industries and helps speed up work processes. However, not all manufacturers are efficient with new technologies. To prepare your manufacturing business for changes and reshape it according to trends, engineers and managers must be open to technological capabilities. They must prepare their employees for the changes and the process of introducing and adapting to technology.

2.3.1 Classic maintenance methods

For many companies, managing industrial machine maintenance represents a major challenge. However, there are proven methods based on impeccable organization, fluid communication, and efficient technological tools. **Figure 5** summarizes the most common classic methods in the field of industrial maintenance management [14].

Figure 4.
Benefits of predictive maintenance.

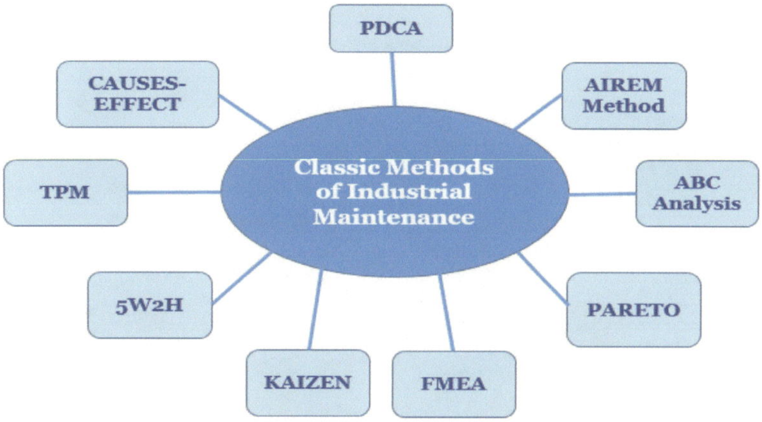

Figure 5.
Classic methods of industrial maintenance.

These methods must be combined with CMMS software for efficient and smooth industrial maintenance of company equipment. The CMMS is considered a data centralization platform, and this maintenance solution makes it possible to manage the activity, edit reports, and monitor indicators in order to make more relevant decisions.

2.3.2 Prognostics and Health Management (PHM) as a new method of maintenance

Global performance requirements are leading manufacturers to strengthen their capacity to anticipate degradation phenomena and breakdowns. Subsequently, Prognosis and Health Management (PHM) solutions are increasingly implemented to complement maintenance activities [15–17]. Maintaining industrial systems in operational condition at a lower cost has become a critical factor. From the concept of PHM to predictive maintenance, it describes the emergence of this discipline, which complements traditional maintenance activities with a more proactive consideration of failures.

The overall principle of Prognostics and Health Management is to transform a set of raw data collected on the monitored equipment into health indicators, whose extrapolation over time makes it possible to define detailed decision support. There are several architectures in the field of Industry 4.0 among those that are developed and the best known are Open System Architecture for Condition-Based-Maintenance [5–9]. An overall view of this PHM architecture is given in **Figure 6**.

This architecture is made up of seven functional layers:

Table 2 describes the seven layers of the OSA/CBM architecture.

The main objective of prognosis is to estimate the Residual Life (RUL) of a system by projecting the evolution of its state of health in the future at an early stage of degradation. Failure prognosis can be carried out using different methods using different modeling, processing, and analysis tools. These methods can be grouped into different categories or approaches such as physical model-based approach, data-driven approach, and experiment-based approach [18, 19].

Figure 6.
OSA/CBM architecture [15–17].

Layer	Layer name	Layer description
L1	Data acquisition	This module provides the system with digital data from sensors or transducers. It covers the different areas of measurement, whether mechanical, electrical, destructive, and tribological.
L2	Signal processing	This module receives signals and data from sensors or, transducers or other signal processors, and performs signal processing through transformation and feature extraction.
L3	Monitoring	This module receives data from signal processing modules and other monitoring modules. It compares data with reference values and must also be able to generate alerts based on predefined thresholds.
L4	Diagnostic	This module determines whether the state of the monitored system or component is degraded or not and identifies the fault responsible for this degradation.
L5	Prognosis	This module relies on data from previous modules, and it predicts the future state of the monitored system and its components by projecting the current state of health of the system into the future.
L6	Help with the decision	This module aims to recommend maintenance actions or other alternatives linked to the management of the system to ensure the continuity of its operation.
L7	Human-machine interface (HMI)	This module builds an Human-Machine Interface (HMI) for the system, and it displays information on the health status of the system as well as alerts received from other previous modules.

Table 2.
The 7 layers of the OSA/CBM architecture.

3. General information on artificial intelligence

Artificial intelligence (AI) corresponds to a set of technologies that makes it possible to simulate intelligence and automatically accomplish tasks of perception, understanding, and decision-making. These techniques particularly involve the use of computer science, electronics, mathematics (notably statistics), neuroscience, and cognitive sciences.

3.1 Definition of artificial intelligence

Historically, work in AI began in the 1950s with the work of Alan TURING. AI became a field of research in the summer of 1956, during the first conference of the pioneers in this discipline, notably John MCCARTHY, Marvin MINSKY, Allen NEWELL, Herbert SIMON, and Donald MICHIE. Before 2000, the limits imposed by calculation and storage capacities did not allow significant progress to be made in the field of AI. We had to wait until the beginning of the 2000s to see the main factors of technological disruption that enabled current advances appear:

- The Internet network and the shared use of data have made it possible to create technologies such as search engines or decentralized and hyperscalable architectures.
- Exponential growth in the quantity of data: The storage space offered for €1 doubles every 14 months.
- Exponential growth in computing capacity: The total amount of data created each year doubles every 2 years.
- Mobility and the development of connected objects which promote access to real-time data flows: In 2020, there will be 50 billion connected objects, which will produce 10% of the total data created (**Figure 7**).

As a result, artificial intelligence [2, 20] has developed very strongly for more than 10 years with an acceleration in the last 5 years, to enable uses such as:

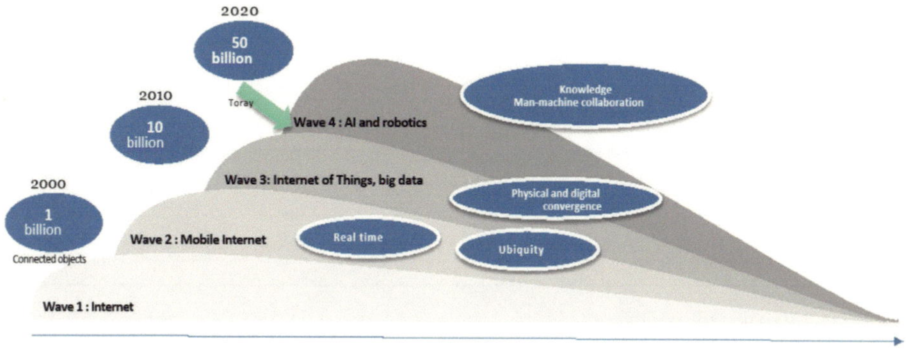

Figure 7.
Chronology of digital developments.

- Visual perception: recognition of an object or description of scenes.
- Understanding written or spoken natural language: automatic translation, automatic production of press articles, sentiment analysis.
- Automatic analysis by "understanding" a query and returning relevant results, even if that result does not contain the query words.
- Autonomous decision-making.

AI currently requires considerable data resources and computing power to learn effectively. Research is now developing techniques to reduce energy consumption and limit the need for data, and other techniques to make it possible to generalize a solution to several uses or to make AI robust in the face of an isolated disruptive event.

Apart from technological aspects, AI also poses new questions in terms of ethics and risk: dependence on automation, misuse, or error linked to "contaminated" data, impact on private life, etc. These questions require us to think about defining the trust framework to be put in place as these technologies develop. Even though artificial intelligence is mainly associated with a mathematical discipline and algorithmic techniques, it also includes other bricks to meet full use. The main building blocks of an AI system are shown in **Figure 8**.

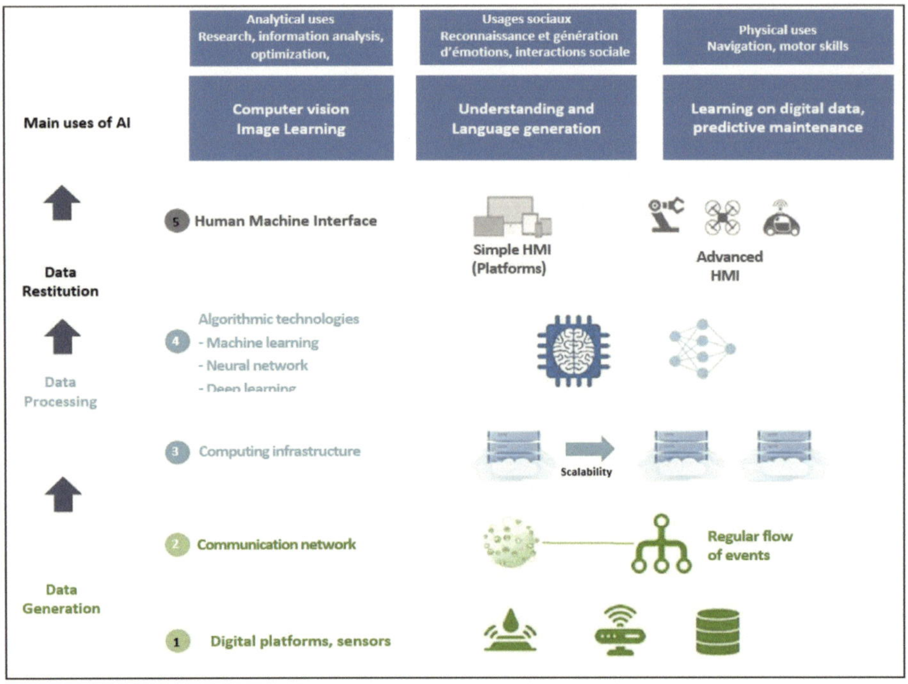

Figure 8.
AI technological building blocks.

3.2 Algorithmic technologies

This section presents the algorithmic techniques of artificial intelligence, particularly the techniques of Machine Learning (see **Figure 9**). These algorithmic techniques make it possible to develop analysis, prediction, and decision-making capabilities and to intelligently and constantly adapt to situations based on data already acquired and currently being acquired [20].

The Machine Learning [13, 24–26] process shown in **Figure 10** begins by defining a cognitive task to automate. This can be a perception, comprehension, or decision task (by level of increasing complexity). We identify the flow data likely to respond to the problem posed. The problem to be solved and the type of data available often determine the type of AI algorithm that we will be able to use.

Furthermore, the quality of automation invariably relies on the following triptych: data, expertise in Machine Learning to model, and business expertise to define usage and interpret the results.

3.3 AI algorithms applied to maintenance

Artificial intelligence offers tools that are completely decoupled from the structure of the system, not requiring prior modeling of the latter and allowing real-time monitoring of its evolution, and its tools use several algorithms, among others, those applied for maintenance [22].

3.3.1 Methods based on behavioral models

There are two main approaches for building these models: finite state automata and Petri nets.

3.3.1.1 Finite state automata

See **Figure 11**.

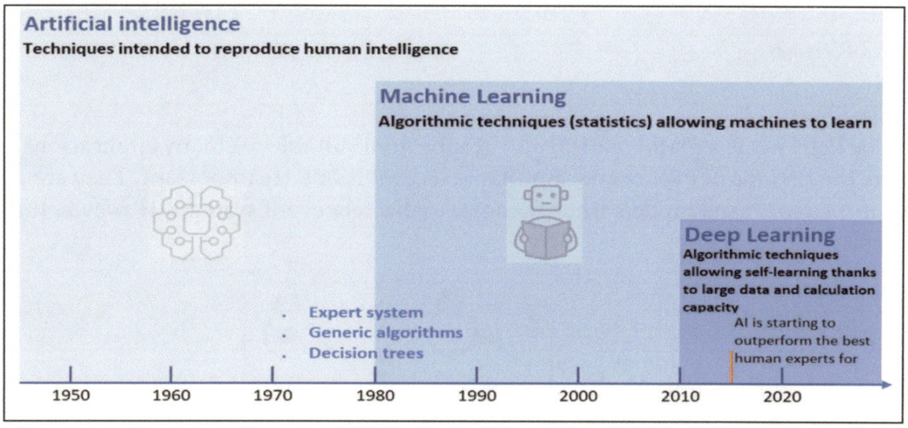

Figure 9.
Timeline of artificial intelligence algorithmic techniques [21–23].

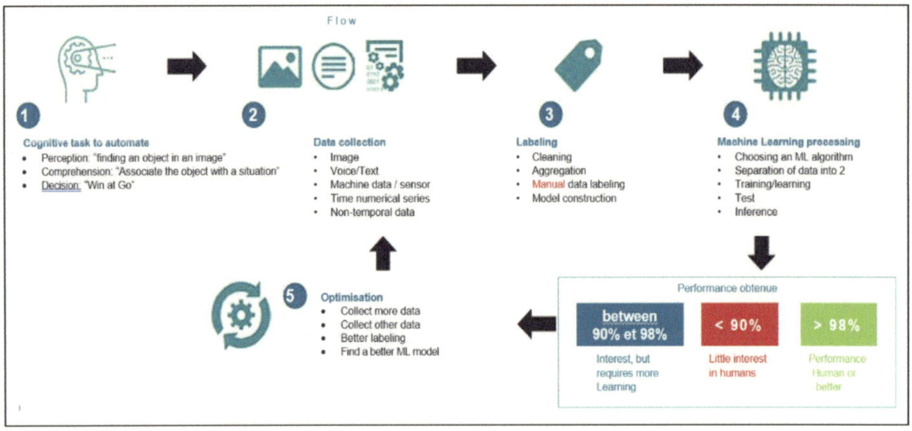

Figure 10.
Machine learning process [23, 27].

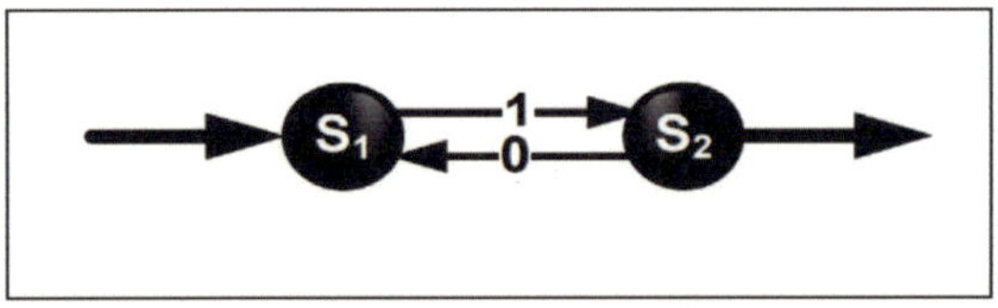

Figure 11.
Very simple example of a finite automaton.

They make it possible to directly model the operation of the system, thanks to a global automaton obtained by composing elementary automata corresponding to local systems (system components). This representation is therefore directly adapted to simulation and detection. However, there are systems for which this representation is also used for diagnosis. A method presented in Refs. [14, 15] is characterized by two steps to perform the diagnosis.

3.3.1.2 Petri nets

See **Figure 12**.

The Petri net is a mathematical and graphic tool suitable for many applications where the notions of events and simultaneous evolutions are important. They are one of the most used models when it comes to discrete event systems. However, they

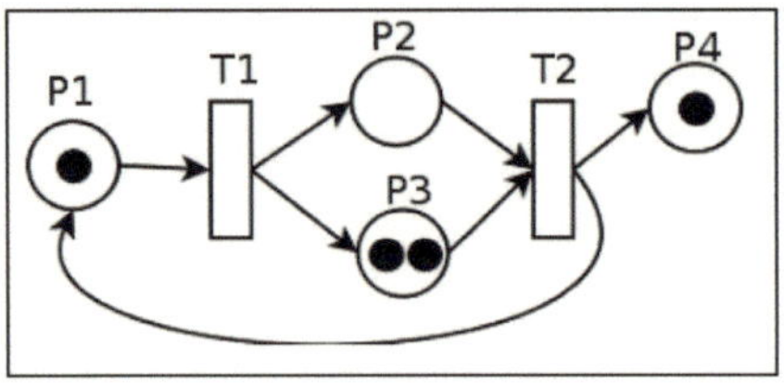

Figure 12.
Example of a petri net.

have been enriched in several aspects (timed, stochastic, fuzzy RdP), so as to better account for the dynamics of discrete event systems. Used initially as generating models, they allow simulation to be carried out as well as detection with a view to use in system diagnostics. In this context, Petri nets can be qualified as models of good functioning. In Ref. [15], backward chaining type reasoning on Petri nets is defined.

3.3.1.3 Other formalisms

There are also other formalisms to be linked to methods based on behavioral models such as qualitative physics models which make it possible to obtain a model by abstraction from the numerical model [12] or approaches in classical or linear logic (also used with Petri nets) [2].

Finite state automata and Petri nets therefore constitute relatively well-suited tools for constructing detection mechanisms when the normal operation of the system is described by these formalisms. On the other hand, their uses in diagnosis are still limited. For automata, the main difficulties are linked to the large size of the state space, which therefore lead to problems with memory and speed of diagnostic execution. As highlighted in Ref. [2], Petri nets constitute a powerful modeling tool and can be considered as a tool for describing the knowledge necessary for diagnosis.

3.3.2 Pattern recognition methods for surveillance

These methods assume that no model is available to describe cause-and-effect relationships. The only knowledge is based on human expertise supported by solid feedback [24]. Most of these methods are based on artificial intelligence with, in particular, tools such as expert systems, statistical tools (pattern recognition), case-based reasoning (CAR), neural networks, fuzzy logic, and neuro-fuzzy networks.

3.3.2.1 Expert systems

See **Figure 13**.

An expert system [23] is software that reproduces the behavior of a human expert performing an intellectual task in a specific domain. It is composed of two independent parts:

- a knowledge base, itself composed of a rule base which models the knowledge of the domain considered and a fact base containing the information concerning the case that we are currently processing.

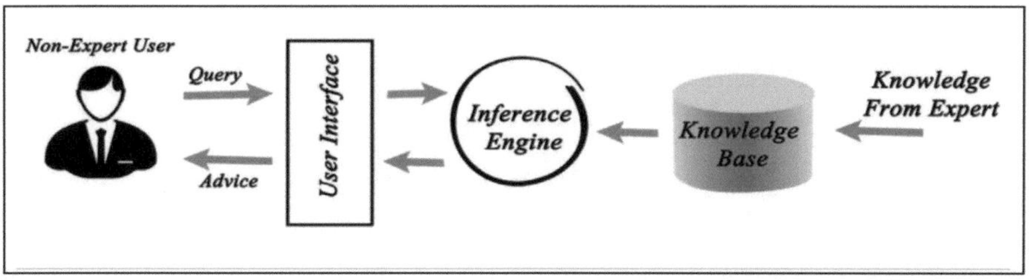

Figure 13.
Expert system.

- an inference engine capable of reasoning from the information contained in the knowledge base, of making deductions, etc.

3.3.2.2 Statistical tools for pattern recognition

The first technique presented is a classic probability-based discrimination technique. This technique may prove insufficient because it assumes *a priori* knowledge of all operating states and does not consider the evolution of the system [24].

3.3.2.3 Case-Based Reasoning—CBR

Case-Based Reasoning (CBR) is a recent approach to solving and learning problems. It corresponds to solving a new problem by remembering a previous similar situation and reusing information and knowledge from that situation [25]. The principle of operation of the method consists of storing previous experiences (cases) in memory to solve a new problem [21, 25, 26]:

- find the experience similar to the new problem in memory,

- reuse this experience in the context of the new situation (completely, partially, or by adapting it according to differences),

- memorize the new experience in memory (learning).

3.3.2.4 Case structure

The structure of the cases will depend on the areas of use and the tasks to be accomplished. Adapted to the diagnosis, the structure of the cases is therefore as follows:

- Problem ↔ symptoms (description of the diagnostic situation)

- Solution ↔ origins (several possible origins)

- Conclusion ↔ actions (maintenance strategy)

Some more recent work has been developed on the use of CBR for diagnosis [28]. In conclusion, CBR constitutes a technique for solving problems based on experience, and therefore relatively well-suited to diagnostic problems for which the notion of experience is relatively important.

3.3.3 Pattern recognition using neural networks

Neural networks are tools capable of performing perception, classification, and prediction operations. Their operation is based on the operating principles of biological neurons. Their main advantage over other tools is their ability to learn and generalize their knowledge to unknown inputs. One of the qualities of this type of tool is its suitability for the development of modern monitoring systems, capable of adapting to a complex system with multiple reconfigurations. Neural networks can also be implemented in electronic circuits, thus offering the possibility of real-time processing.

Their use is mainly guided by their following properties:

- learning ability,
- generalization ability,
- parallelism in processing (speed of processing),
- adapted to the nonlinearities of systems.

Each neuron performs a simple function (linear function, piecewise linear function, threshold function, sigmoid, Gaussian), with the global properties of the tool emerging from its structure. All the characteristics of neural networks are exploited through the main property of neural networks, which is learning. Indeed, learning mechanisms are at the origin of the problem-solving capabilities of neural networks. This learning makes it possible to configure the synaptic weights as well as the activation functions to adopt a desired behavior. Two types of learning are used: supervised learning and unsupervised learning [23, 27].

3.3.4 Pattern recognition using fuzzy logic

Fuzzy logic makes it possible to formalize the representation and processing of imprecise or approximate knowledge. It offers the possibility of dealing with highly complex systems in which, for example, human factors are present. It intervenes in the manipulation of imperfect knowledge. In these various applications, the use of fuzzy logic is quite natural, insofar as it makes it possible to deal with imprecision, uncertainty, and incompleteness linked to domain knowledge. In addition, fuzzy logic gives them the ability to be used in prognosis [28].

Neuro-fuzzy networks were born from the association of neural networks with fuzzy logic, so as to take advantage of the benefits of each of these two techniques. The main property of neuro-fuzzy networks is their ability to process digital and symbolic knowledge of a system in the same tool [28]. They, therefore, make it possible to exploit the learning capabilities of neural networks on the one hand and the reasoning capabilities of fuzzy logic on the other hand. Different combinations of these two artificial intelligence techniques exist and highlight different properties. The following combinations can be identified [28]: Neural fuzzy network, Neural/blur system simultaneously, Cooperative neuro-fuzzy models, and Hybrid neuro-fuzzy models.

Neuro-fuzzy structures for modeling, prediction, control, or diagnosis can be produced by a wide variety of architectures for the same type of given combination. For example, in Ref. [23], a use of a neuro-fuzzy system Recurrent Self-Adaptive Neuro-Fuzzy Inference System (R-SANFIS) for controlling an autonomous underwater vehicle. Another use of neuro-fuzzy networks is presented in Refs. [23, 27], where the NEuro Fuzzy function apPROXimator (NEFPROX) architecture is used for function approximation.

In diagnostic applications, we mainly find hybrid neuro-fuzzy models, for which neural network and fuzzy system are combined homogeneously.

3.3.5 Methods based on explanatory models

These methods are mainly based on representing the relationships between different fault states and their (possibly observable) effects. They are therefore based on a deep analysis of the system, so as to have sufficient knowledge to express its cause-and-effect relationships. The models thus obtained allow—for some—an abductive approach, which consists of going back to the causes of breakdowns from observations corresponding to the symptoms. Several artificial intelligence tools allow such formalization of the knowledge available on a system. These include causal graphs and contextual graphs, techniques which are also joined by approaches based on fuzzy logic or Petri nets.

3.3.5.1 Causal graphs

The exploitation of causal knowledge is quite natural for diagnosis. Indeed, a "dysfunction" can be quite simply described by the relationships associating its causes with its observable manifestations. Causal graphs constitute a formalism well-suited to the representation of these causal links. In diagnostic use, they make it possible to express the causal sequences governing the operation of the system to be monitored in the event of a breakdown [29].

The causal graph represents a particularly interesting tool for diagnosis in the sense that it can provide justification for the diagnosis proposed by the system through the causal path followed in the graph. Additionally, abductive diagnosis algorithms make it possible, from the observation of symptoms, to search for a set of possible causes that explain the observations through causal relationships. Finally, the introduction of temporal constraints, contradictory effects, and the consideration of interactions between failures more accurately reflect the physical reality of the system to be diagnosed.

3.3.5.2 Contextual graphs

Contextual graphs therefore appear to be a tool suitable for modeling activities involving a procedure/practice duality. They are therefore applicable in areas where an interpretation or adaptation of general rules is necessary to consider the richness of the real context of application. For diagnosis, they will be applied in areas where the causes of system failures are strongly linked to the context in which the failure occurred. In the context of a supervision application, they could be applied in cases where the context takes an important place in the link between fault diagnosis and recovery actions [30, 31].

4. Contribution of AI to maintenance

Thanks to artificial intelligence, all data collected by the sensors are analyzed in real time to find relationships between historical data and current readings, but also to alert technicians in the event of a risk of failure. Companies are taking a more proactive approach to their maintenance strategy. To do this, they rely on data from IoT sensors linked to CMMS software to monitor anomalies and use predictive models to request human intervention. This strategy notably avoids shutting down a production line when it is not useful and reduces corrective maintenance which is expensive and disrupts the activity [32, 33].

To make the transition to 4.0 maintenance, it is important to proceed gradually according to the starting level, the internal resources, and the appetite for digital technology. Indeed, each technological development involves profound changes that affect organization, teams, and technical infrastructure [33].

4.1 Toward maintenance 4.0

Maintenance 4.0 makes it possible to move from classic corrective maintenance to more intelligent predictive maintenance. Thanks to digital technologies, it is no longer possible to react to breakdowns but to anticipate them and deal with them before they occur. It allows maintenance costs to be optimized and product quality to be improved (see **Figure 14**) [34, 35].

4.1.1 Technological bases

The technological bases of predictive maintenance are based on advanced technologies and data processing capabilities [31]. Below are the essential elements that constitute predictive maintenance:

4.1.1.1 Sensors and data collection

Predictive maintenance relies on sensors integrated into machines and installations. The sensors continuously collect data on the condition of the machines, such as vibrations, temperature, pressure, flow rates, and much more. They collect data in real time and transmit them to data processing platforms (see **Figure 15**).

The form of communication used until now, between sensors, control units and the process control, production, and business level, constitutes a closed system. Data

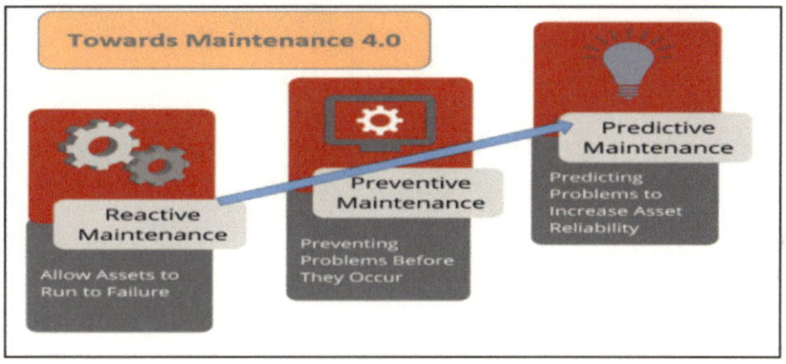

Figure 14.
Toward maintenance 4.0.

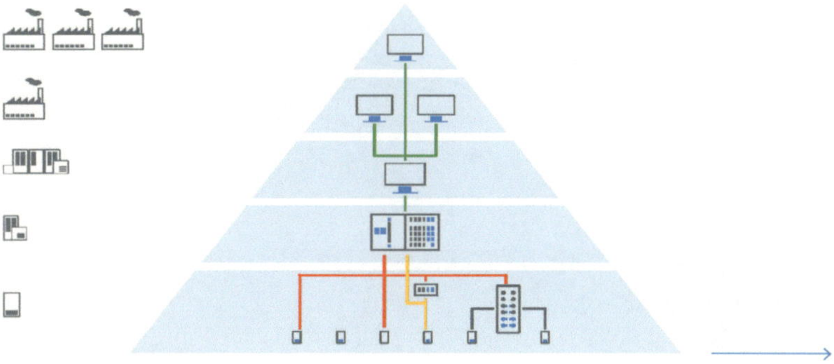

Figure 15.
Current communication levels.

are transmitted from field devices, namely sensors and actuators, to the programmable logic controller (PLC) (see **Figure 16**).

Decentralized computing power converts data into information directly in the sensor. Decisions will be decentralized. Process, production, and business-relevant information will be transmitted directly to the Ethernet and the cloud (see **Figure 17**).

In the future, the cloud will gain ground in general process management. But core computing power will increasingly shift to the edge. The sensors convert the collected data into information which is then processed in the Ethernet or cloud for the next process.

4.1.1.2 IoT (Internet of Things)

The Internet of Things [23] plays a crucial role in enabling the networking of sensors, machines, and installations and the transparent transmission of data to central platforms [12, 18, 36, 37] and processing of data or cloud-based systems is thus guaranteed (see **Figure 18**).

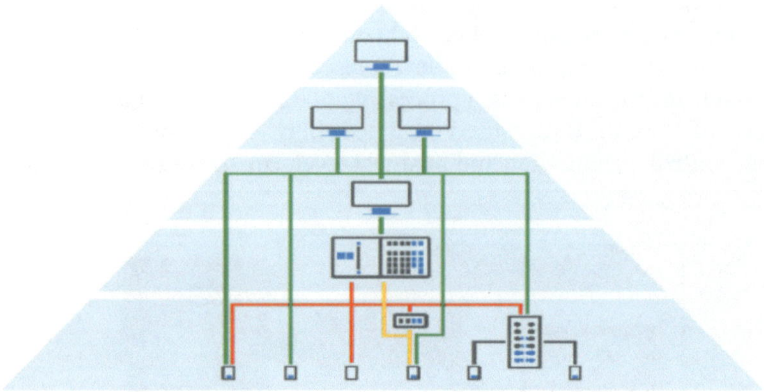

Figure 16.
Communication levels in the age of Industry 4.0.

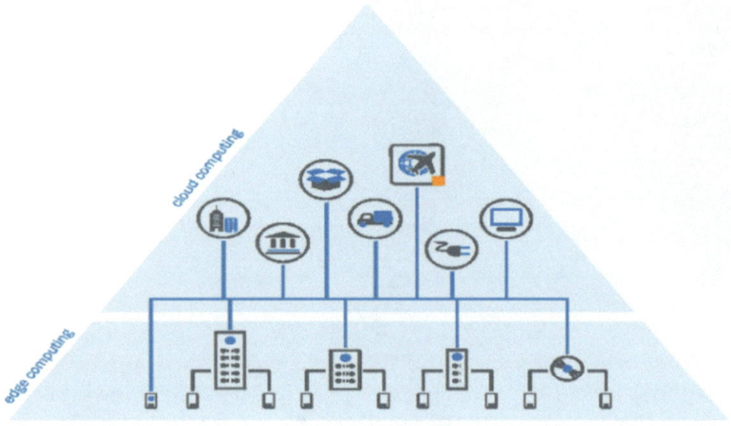

Figure 17.
Connected information.

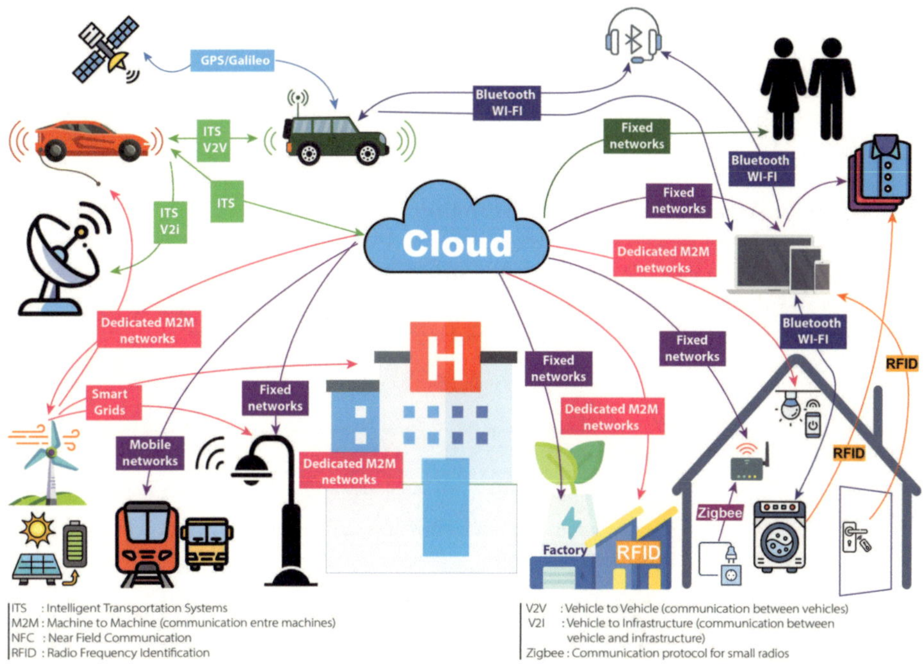

Figure 18.
Connected objects and the highlight [34].

The Internet of Things (IoT) is radically transforming the way businesses approach managing their data and IT systems. With each passing day, users generate a massive amount of data, reaching up to 2.5 quintillion bytes, creating a significant challenge to stay relevant in an ever-changing digital environment. However, IoT is emerging as a key solution to making sense of this abundance of data by introducing automation where it is needed.

The Workflow Management Software is specifically designed to aggregate data from these connected devices, bringing significant improvements in both the professional and personal spheres:

- Data processing and analysis
- Algorithms and models
- Predictive analytics
- Integration into business systems
- Continuous learning
- Data entry and processing
- Data analysis and modeling

Data analysis and modeling allow businesses to identify patterns and anomalies in their data and respond quickly to potential issues before unplanned downtime occurs. This helps to improve plant availability and reduce maintenance costs.

4.1.2 Implementation of predictive maintenance

Implementing predictive maintenance requires a well-thought-out strategy and approach structured. Here are the steps businesses should take to implement maintenance [31, 38, 39] predictive (see **Table 3**).

4.2 AI contribution statistics for maintenance

Global industry, particularly the aeronautics sector and the rail transport sector, is booming thanks to Maintenance 4.0 linked to technological changes, notably the integration of IoT equipment and associated data [33, 40].

Steps	Description of steps
Define goals and requirements	• Define clear objectives for the implementation of predictive maintenance. • Identify the specific machines or installations for which predictive maintenance must be implemented.
Identify data sources	• Identify relevant data sources and sensors needed for machine health monitoring. • Ensure that data can be collected and transmitted in real time to a central platform or system.
Set up a data infrastructure	• Establish a robust data infrastructure that enables data collection, storage, and processing. • Take security and data protection regulations into account when processing data.
Ensuring data quality	Monitor and maintain the quality of collected data to ensure it is suitable for analysis. This may include data cleaning and noise removal.
Choosing analysis and modeling techniques	Decide which analysis and modeling techniques best suit your needs. This could be machine learning, statistical models, or a combination of both.
Model development and training	Develop and train models based on historical data. Use these models to monitor machine health in real-time.
Set thresholds and alarms	Set thresholds and criteria that determine when alarms or notifications trigger. This helps identify problems proactively.
Integration into existing processes	• Integrate predictive maintenance into your existing maintenance and operations processes. • Ensure that maintenance personnel can effectively use information and alarms.
Training and awareness	• Train your team in the use of predictive maintenance tools and systems. • Raise your employees' awareness of the importance of the new strategy and how it helps improve efficiency.
Monitoring and optimization	• Establish a continuous monitoring system to ensure that the predictive maintenance strategy is effective. • Continuously optimize models and algorithms to improve prediction accuracy.
Measuring success	Define clear key performance indicators (KPIs) to measure implementation success. These can include indicators such as reduced downtime, reduced maintenance costs, and increased plant availability.
Data ethics and data protection	Data ethics and data protection must be respected throughout the process, especially when it comes to data collection and storage.

Table 3.
Steps for implementing predictive maintenance.

For many industries, predictive maintenance remains a competitive and profitable solution in terms of investment. They have started to reap the benefits thanks to the integration of new AI technologies, which have changed and continue to change the business landscape and industrial system technology. **Figure 19** shows the contribution shares of predictive maintenance by sector of activity:

Companies that have fully adopted AI-powered software and are committed to adopting AI highlight several benefits that show how AI has changed the game. According to "Grand View Research," the global artificial intelligence market is valued at $136 billion.

- Growth of 1400% is forecast for the next 7 years.
- In 2030, this market is expected to be valued at more than $1.81 trillion.
- In 2018, the market generated only $10 billion annually.
- In 2025, 97 million people will work in the field of artificial intelligence.

Figure 20 presents the result of company statistics carried out by "MIT Sloan Management" and "Sales Forces" in answering the question on the use of artificial intelligence and its applications, and their answers were as follows:

4.2.1 Impact of AI on the industrial sector

AI technology is slowly penetrating all spheres of our lives and is being incorporated into all kinds of devices and software. In 2023, AI is everywhere around us: in the industrial sector thanks to maintenance 4.0, autonomous cars are already much more than a simple element of a science fiction film, virtual assistants are efficient and credible, and countless software uses the technique of machine learning [41, 42]. As technology advances, AI capabilities increase and become not only more practical for the industry but also a new standard.

According to statistics carried out by "Passport-Photo.online" shows that 78% of IT managers are already using AI or planning to do so to automate workflow. After all, its abilities, at the very least, make life significantly easier in handling tasks that would otherwise be time-consuming and arduous for the human mind. Data management

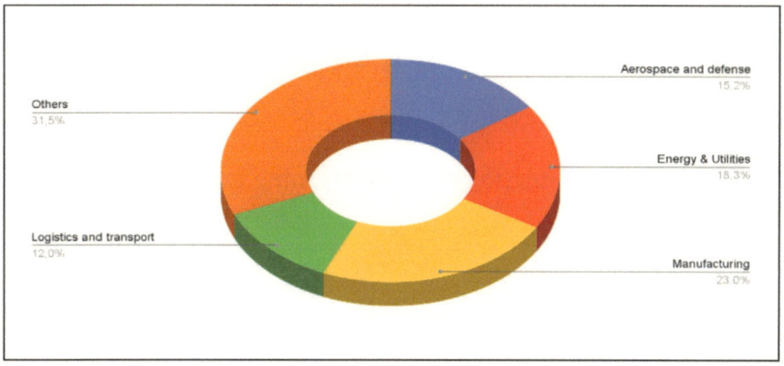

Figure 19.
Contribution of predictive maintenance by sector of activity. Source: JDN Journal Du Net/Predictive maintenance market, Global forecast to 2021.

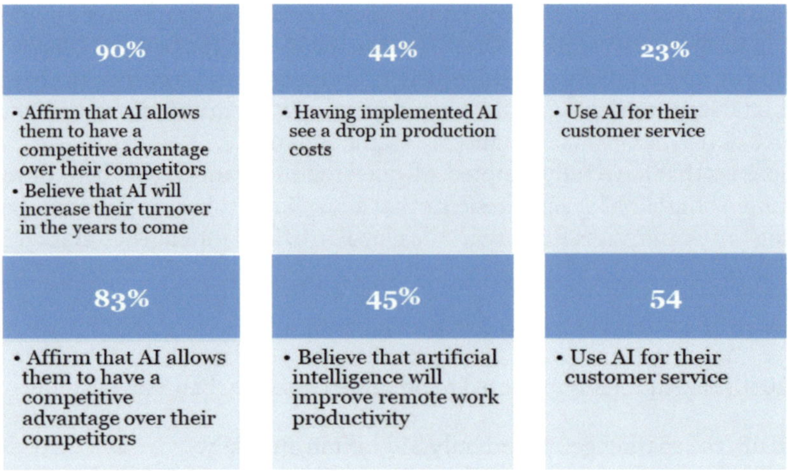

Figure 20.
Statistics result.

alone is one piece that confirms the AI is truly a game-changer in how businesses operate and take advantage of new possibilities, not to mention machine learning [43–45].

4.2.2 Benefits of artificial intelligence

In recent years, considering the rise of AI and its omnipresence in daily life, we can affirm that this development would not have taken place if there had been only a handful of advantages of its implementation. The latest AI statistics show that the path to success as a business owner is to immerse yourself in AI's capabilities and exploit them before the competition does. It is particularly interesting to note that the first benefit mentioned is improving the customer experience, something that many observers believe should be at the top of a company's priorities. The same study shows that most people (63%) believe that AI will help solve the problems of modern society and help us live more fulfilling lives. This is another very positive and remarkable result that counters the fear that AI will become a problem in the future [46]. Integrating AI into predictive maintenance has many benefits:

- Optimized equipment or system availability
- Getting closer to operational excellence through "zero breakdown"
- Reduced maintenance costs
- Optimization of the production chain by planning machine maintenance operations at the right time
- Reduction in breakdowns
- Improvement in the supply of spare parts to limit excess stock.
- Increased lifespan of operating assets
- Increased customer satisfaction through quality service on time

DOI: http://dx.doi.org/10.5772/intechopen.1005280

4.3 Artificial intelligence trends

By following a structured approach to implementation and leveraging the right technology, processes, and expertise, organizations can realize significant benefits in terms of equipment reliability, maintenance efficiency, and operational performance [47–49]. Since AI has the capacity to analyze considerable quantities of data that would be impossible to analyze manually, its integration into industrial systems allows companies to have advanced tools to optimize the reliability and efficiency of different manufacturing operations maintenance, such as planning interventions, purchasing spare parts in advance, or wasting raw materials and energy. The primary objective is always to avoid costly production interruptions. Data analysis is truly a key element for real-time detection of technical problems. By using AI to analyze data from sensors and machines, anomalies and potential issues are quickly identified. This allows you to react as quickly as possible and reduce operational interruptions [48].

4.3.1 Success stories of the contribution of AI to predictive maintenance

AI is a powerful tool that can significantly increase the effectiveness of predictive maintenance by allowing companies to monitor the health of their equipment, detect possible future failures, and quickly notify maintenance teams of a check to be carried out so that they can plan interventions and avoid possible costly shutdowns.

The reduction in corrective and curative maintenance generates real financial gain for companies. By using information from sensors and equipment performance, AI can therefore help identify equipment that requires more frequent or intensive intervention [48–50].

As follows, some case studies and success stories are given [46, 48, 51].

4.3.1.1 General Electric (GE) Aviation

GE Aviation implemented a predictive maintenance program called “Prognostic Health Management” (PHM) for aircraft engines. By equipping engines with sensors to collect real-time performance data, GE Aviation can monitor engine health, predict potential failures, and schedule maintenance proactively. As a result, airlines can avoid unscheduled downtime, reduce maintenance costs, and optimize fleet operations. For example, GE Aviation’s PHM system helped Cathay Pacific Airways reduce engine related delays by 35% and decrease maintenance costs by 10%.

4.3.1.2 Schneider Electric

Schneider Electric, a global leader in energy management and automation solutions, implemented predictive maintenance for its electrical distribution equipment. By analyzing data from sensors embedded in switchgear, circuit breakers, and transformers, Schneider Electric can detect early signs of equipment degradation or malfunction and schedule maintenance proactively. As a result, customers experience increased equipment reliability, reduced downtime, and improved safety. For example, a manufacturing facility in Europe reduced equipment downtime by 20% and achieved a return on investment (ROI) within 6 months of implementing Schneider Electric’s predictive maintenance solution.

4.3.1.3 Rio Tinto

Rio Tinto, a multinational mining corporation, implemented predictive maintenance for its fleet of autonomous haul trucks used in mining operations. By deploying sensors and predictive analytics software, Rio Tinto can monitor truck performance, identify potential issues, and schedule maintenance proactively to avoid costly breakdowns. As a result, Rio Tinto has increased equipment availability, improved productivity, and reduced maintenance costs. For example, predictive maintenance helped Rio Tinto achieve a 10% increase in truck availability and a 15% reduction in maintenance costs at its Pilbara iron ore mine in Australia.

4.3.1.4 Siemens Gamesa Renewable Energy

Siemens Gamesa Renewable Energy [46, 51], a leading manufacturer of wind turbines, implemented predictive maintenance for its wind turbine fleet to optimize performance and reduce maintenance costs. By analyzing data from sensors embedded in turbines, Siemens Gamesa can detect early signs of component wear, identify potential failures, and schedule maintenance proactively. As a result, wind farm operators experience increased turbine availability, higher energy production, and reduced operational costs. For example, predictive maintenance helped a wind farm operator in Europe achieve a 20% reduction in maintenance costs and a 10% increase in energy production.

4.3.1.5 Pacific Gas and Electric Company (PG&E)

PG&E, a utility company serving millions of customers in California, implemented predictive maintenance for its electrical distribution network to improve reliability and reduce outage duration. By analyzing data from smart meters, sensors, and weather forecasts, PG&E can identify potential equipment failures, prioritize maintenance activities, and deploy crews proactively to restore service quickly in the event of an outage. As a result, PG&E has achieved significant improvements in grid reliability, customer satisfaction, and operational efficiency. For example, PG&E reduced outage frequency by 15% and outage duration by 20% within the first year of implementing predictive maintenance.

These case studies demonstrate the diverse applications and benefits of predictive maintenance across different industries, including aviation, manufacturing, mining, renewable energy, and utilities. By leveraging advanced technologies and data analytics capabilities, organizations can optimize maintenance practices, enhance asset reliability, and drive operational excellence, ultimately delivering value to customers and stakeholders [46, 51].

4.3.2 Trends in industrial maintenance with the exploitation of AI

Artificial intelligence technologies will be increasingly used in the field of maintenance and Industry 4.0. One of the major axes for the application of AI in industry is the quality and quantity of available information. For AI to work effectively, it is necessary to have high-quality and sufficient information to train and power the models and algorithms.

This is particularly one of the challenges with CMMS software, which must allow maintenance services to easily collect and store important information. AI is of great importance to industries, and the maintenance and field intervention sectors are obviously no exception. Indeed, AI makes it possible to optimize processes to make more efficient decisions based on clear and precise data. The proof, according to a report recently released by Bitkom statistics, is that half of the companies specializing in the sector believe that the gain in productivity is the main advantage associated with the use of artificial intelligence. It is a fact: AI can optimize the management and maintenance of industrial systems.

1. Advanced machine learning techniques: future advancements in machine learning techniques are expected to drive improvements in predictive maintenance capabilities. Deep learning architectures, such as deep neural networks (DNNs), convolutional neural networks (CNNs), and recurrent neural networks (RNNs), will continue to be explored for their potential to extract complex patterns and relationships from sensor data, enabling more accurate fault detection and diagnosis.

2. Edge computing and IoT integration: the integration of edge computing and Internet of Things (IoT) technologies will enable real-time processing and analysis of sensor data at the network edge, reducing latency and bandwidth requirements. Edge-based predictive maintenance solutions will allow organizations to perform data analytics and decision-making closer to the source of data generation, enabling faster response to equipment anomalies and minimizing reliance on centralized cloud infrastructure.

3. Digital twins and simulation modeling: digital twin technology, which creates virtual replicas or models of physical assets, will play a key role in predictive maintenance by enabling simulation-based modeling and predictive analytics. By coupling real-time data from sensors with virtual representations of equipment, organizations can simulate different maintenance scenarios, predict the impact of interventions, and optimize maintenance strategies to maximize asset performance and reliability.

4. Explainable AI and interpretability: as AI-based predictive maintenance solutions become more pervasive, there will be a growing emphasis on explainability and interpretability of predictive models. Explainable AI techniques, such as model-agnostic methods, feature importance analysis, and visualization tools, will enable stakeholders to understand the rationale behind model predictions, build trust in AI systems, and facilitate human-in-the-loop decision-making.

5. Predictive Maintenance as a Service (PMaaS): the rise of cloud computing and subscription-based models will pave the way for Predictive Maintenance as a Service (PMaaS) offerings. PMaaS providers will offer scalable, cost-effective predictive maintenance solutions hosted on cloud platforms, allowing organizations to access advanced analytics capabilities, predictive models, and expertise without the need for extensive infrastructure investment or in-house data science resources.

6. Autonomous maintenance systems: advancements in artificial intelligence, robotics, and autonomous systems will enable the development of autonomous maintenance systems capable of self-diagnosis, self-repair, and self-optimization. These systems will leverage AI-based algorithms to continuously monitor

equipment health, detect anomalies, and perform maintenance tasks autonomously, reducing the need for human intervention and improving overall operational efficiency.

7. Integration with Industry 4.0 initiatives: predictive maintenance will become an integral component of Industry 4.0 initiatives aimed at digitizing and optimizing manufacturing processes. Integration with technologies such as cyber-physical systems, additive manufacturing, and augmented reality will enable seamless data exchange, real-time monitoring, and adaptive maintenance strategies, leading to increased productivity, reduced downtime, and enhanced competitiveness.

8. Ethical and responsible AI practices: with the proliferation of AI-based predictive maintenance solutions, there will be a growing emphasis on ethical and responsible AI practices. Organizations will need to address concerns related to data privacy, bias, fairness, and accountability in AI systems to ensure that predictive maintenance initiatives are deployed in a socially responsible manner and aligned with ethical principles and regulatory requirements.

In summary, the future of AI-based predictive maintenance will be characterized by advancements in machine learning techniques, edge computing, digital twin technology, and autonomous systems, enabling organizations to achieve higher levels of asset reliability, operational efficiency, and sustainability. By embracing emerging trends and leveraging innovative technologies, organizations can stay ahead of the curve and unlock new opportunities for predictive maintenance optimization and innovation.

5. Conclusion

In this chapter, we have covered the growing integration of artificial intelligence in the field of mechanical maintenance, which offers significant benefits in terms of operational efficiency, cost reduction, and improved equipment reliability. Using techniques such as machine learning and predictive analytics, AI-based systems can anticipate failures, optimize maintenance plans, and extend the useful life of machines.

We explored how AI can be used to optimize the operating parameters of industrial systems equipment. This approach plays an increasingly important role in improving the management of intervention methods and increasing the efficiency of operations starting from fault detection to operations planning, including in-depth data analysis to propose controls and plan maintenance interventions at the best time.

We discussed the contribution of AI to industrial maintenance, thanks to connected devices and the exploitation of new generations of CMMS, moving from classic reactive maintenance to more intelligent and proactive maintenance which will offer real benefits to industrial companies in terms of the performance of their production systems.

We also covered the different basic notions of industrial maintenance through its concepts, technologies, and methods used. However, to take full advantage of these methods and technologies, it is essential to develop specialized skills, ensure data quality, and maintain a balance between automation and human intervention.

In our future work and to strengthen the decision-making reliability of our predictive maintenance system, we will focus on a case study by proposing a method of diagnosis and prognosis of the imbalance fault of a rotating machine using the techniques and algorithms of artificial intelligence and machine learning by highlighting their advantages over traditional techniques of vibration analysis and diagnosis, mainly temporal analysis and frequency analysis of signals.

Ultimately, AI represents a powerful tool to modernize and improve maintenance processes in the mechanical sector, thus helping to increase the productivity and competitiveness of companies.

Conflict of interest

The authors declare no conflict of interest.

Author details

Mohamed El Khaili*, Mohamed Rafik, Redouane Fila and Abdelmajid Farid
IESI Lab ENSET, Hassan 2nd University of Casablanca, Mohammedia, Morocco

*Address all correspondence to: elkhaili@enset-media.ac.ma

References

[1] Thoben KD, Wiesner S, Wuest T. "Industry 4.0" and smart manufacturing—A review of research issues and application examples. International Journal of Automotive Technology. 2017;**11**:4-19. DOI: 10.20965/ijat.2017.p0004

[2] Bachri K, Anggoro B, Sumari A, Ahmad A. Cognitive artificial intelligence method for interpreting transformer condition based on maintenance data. Advances in Science, Technology and Engineering Systems Journal. 2017;**2**(3):1137-1146

[3] Fila R, El Khaili M, Mestari M. Cloud computing for industrial predictive maintenance based on prognostics and health management. Procedia Computer Science. 2020;**177**:631-638. ISSN 1877-0509

[4] Ayar H, Nhaila H, Khaili ME. Modeling and simulation of power transformer breakdowns to implement predictive maintenance. In: 2023 3rd International Conference on Innovative Research in Applied Science, Engineering and Technology (IRASET). Mohammedia, Morocco: IEEE Xplore Digital Library. 2023. pp. 1-8. DOI: 10.1109/IRASET57153.2023.10152941

[5] Kaiser KA, Gebrael NZ. Predictive maintenance management using sensor-based degradation models. IEEE Transactions on System, Man, Cybern Part A: Systems Humans. 2009;**39**:840-849. DOI: 10.1109/TSMCA.2009.2016429

[6] Chen C, Lu N, Jiang B, Wang C. A risk-averse remaining useful life estimation for predictive maintenance. IEEE/CAA Journal of Automatica Sinica. 2021;**8**(2):412-422. DOI: 10.1109/JAS.2021.1003835

[7] Khan M, Ahmad A, Sobieczky F, Pichler M, Moser BA, Bukovský I. A systematic mapping study of predictive maintenance in SMEs. IEEE Access. 2022;**10**:88738-88749. DOI: 10.1109/ACCESS.2022.3200694

[8] Keller R, Stohr A, Fridgen G, Lockl J, Rieger A. Affordance experimentation-actualization theory in artificial intelligence research—A predictive maintenance story. In: ICIS 2019 Proceedings. Association for Information Systems. 2019. p. 1. ISBN 978-0-9966831-9-7

[9] Keleko AT, Kamsu-Foguem B, Ngouna RH, et al. Artificial intelligence and real-time predictive maintenance in industry 4.0: A bibliometric analysis. AI and Ethics. 2022;**2**:553-577. DOI: 10.1007/s43681-021-00132-6

[10] Ahmed I, Jeon G, Piccialli F. From artificial intelligence to explainable artificial intelligence in industry 4.0: A survey on what, how, and where. IEEE Transactions on Industrial Informatics. 2022;**18**(8):5031-5042

[11] Girit O, Atakok G, Ersoy S. Data analysis for predictive maintenance of servo motors. Shock and Vibration. 2020;**2020**:1-10

[12] Koh XM. Development of Predictive Maintenance Data Analytic for Internet of Things (IoT) Implementation. 2017

[13] Chilukuri SK, Challa NP, Mohan JS, Gokulakrishnan S, Mehta RVK, Suchita AP. Effective predictive maintenance to overcome system failures—A machine learning approach. In: Congress on Intelligent Systems: Proceedings of CIS 2020. Vol. 1. Singapore: Springer; 2021. pp. 341-357.

DOI: 10.1007/978-981-33-6981-8_28 Corpus ID: 236796769

[14] Rasovska I, Chebel-Morello B, Zerhouni N. Classification of the different architectures in maintenance. In: 7th International Congress of Industrial Engineering, Trois-Rivières, Quebec (CANADA). Vol. 23. 2007. pp. 1-12

[15] Camci F, Valentine GS, Navarra K. Methodologies for Integration of PHM Systems with Maintenance Data Tech. Rep., DTIC Document. Vol. 852007. p. 86

[16] Herr N, Nicod J-M, Varnier C, Zerhouni N, Dersin P. Predictive maintenance of moving systems. 2017 Prognostics and System Health Management Conference (PHM-Harbin), Harbin, China; July 2017:1-6. DOI: 10.1109/PHM.2017.8079111

[17] Shukla B, Fan IS, Jennions I. Opportunities for explainable artificial intelligence in aerospace predictive maintenance. PHM Society European Conference. 2020;**5**(1):11

[18] Killeen P, Ding B, Kiringa I, Yeap T. IoT-based predictive maintenance for fleet management. Procedia Computer Science. 2019;**151**:607-613

[19] Shaikh MS, Dalmia K, Bhattacharjee S. Artificial Intelligence Based Industrial Product Maintenance Using IoT. 2020

[20] Tripathi S, Budhraja A. Cloud based sewerage monitoring and predictive maintenance using machine learning. International Research Journal of Engineering and Technology. 2019;**6**(4):2282-2287

[21] Ali MI, Lai NS, Abdulla R. Predictive maintenance of rotational machinery using deep learning. International Journal of Electrical and Computer Engineering (IJECE). 2024;**14**(1):1112-1121

[22] Jittawiriyanukoon C, Srisarkun V. Simulation for predictive maintenance using weighted training algorithms in machine learning. International Journal of Electrical & Computer Engineering. 2022;**12**(3):2088-8708

[23] Dash SK, Raj S, Agarwal R, Mishra J. Automobile predictive maintenance using deep learning. International Journal of Artificial Intelligence and Machine Learning (IJAIML). 2021;**11**(2):1-12. DOI: 10.4018/IJAIML.20210701.oa7

[24] Agostinelli S, Cumo F. Machine learning approach for predictive maintenance IN an advanced building management system. Energy Production and Management in the 21st Century V: The Quest for Sustainable Energy. 2022;**255**:131

[25] Mutsuddi S. Machine learning for predictive maintenance in manufacturing industries. International Journal for Science Technology and Engineering. 2023;**11**(4):1175-1181. DOI: 10.22214/ijraset.2023.50098

[26] Çınar ZM, Abdussalam Nuhu A, Zeeshan Q, Korhan O, Asmael M, Safaei B. Machine learning in predictive maintenance towards sustainable smart manufacturing in industry 4.0. Sustainability. 2020;**12**(19):8211

[27] Deutsch J, He D. Using deep learning-based approach to predict remaining useful life of rotating components. In: IEEE Transactions on Systems, Man, and Cybernetics: Systems. Jan 2018;**48**(1):11-20. DOI: 10.1109/TSMC.2017.2697842

[28] Arunkumar G. AI-based predictive maintenance strategies for electrical equipment and power networks. International Journal of Artificial Intelligence in Electrical Engineering (IJAIEE). 2024;**2**(1):1-13

[29] Raja HA, Kudelina K, Asad B, Vaimann T. Perspective chapter: Fault detection and predictive maintenance of electrical machines. London, UK: IntechOpen; 2022. DOI: 10.5772/intechopen.107167

[30] Frigura-Iliasa M, Simo A, Dzitac S, Frigura-Iliasa FM, Baloi FI. Fuzzy-logic based diagnosis for high voltage equipment predictive maintenance. In: Dzitac I, Dzitac S, Filip F, Kacprzyk J, Manolescu MJ, Oros H, editors. Intelligent Methods in Computing, Communications and Control. ICCCC 2020. Advances in Intelligent Systems and Computing. Vol. 1243. Cham: Springer; 2021. DOI: 10.1007/978-3-030-53651-0_21

[31] Kun P, Yuchen J. Computer prediction model for equipment maintenance using cloud computing and secure data-sharing. Journal of Physics: Conference Series. 2021;**2083**(4). DOI: 10.1088/1742-6596/2083/4/042042

[32] Uzoigwe DO. Implementing predictive maintenance (PdM) programs in food and beverage manufacturing facilities. International Journal of Research and Innovation in Applied Science. 2023;**8**(12):103-118

[33] Terrada L, ElKhaïli M, Ouajji H. Multi-agents system implementation for supply chain management making-decision. Procedia Computer Science. 2020;**177**:624-630. Available from: http://www.sciencedirect.com/science/journal/18770509. DOI: 10.1016/j.procs.2020.10.089

[34] Calabrese F, Regattieri A, Bortolini M. Galizia, FG data-driven fault detection and diagnosis: Challenges and opportunities in real-world scenarios. Applied Science. 2022;**12**:9212. DOI: 10.3390/app12189212

[35] Buabeng A, Simons A, Frempong NK, Ziggah YY. Hybrid intelligent predictive maintenance model for multiclass fault classification. Soft Computing. 2023:1-22. DOI: 10.1007/s00500-023-08993-1

[36] Terrada L, Alloubane A, Bakkoury J, Khaili ME. IoT contribution in supply chain management for enhancing performance indicators. In: 2018 International Conference on Electronics, Control, Optimization and Computer Science, ICECOCS 2018, Art. IEEE; 2018. DOI: 10.1109/ICECOCS.2018.8610517

[37] Christou IT, Kefalakis N, Zalonis A, Soldatos J, Bröchler R. End-to-end industrial IoT platform for actionable predictive maintenance. IFAC-PapersOnLine. 2020;**53**(3):173-178. DOI: 10.1016/j.ifacol.2020.11.028

[38] Gościniak T, Wodarski K. Effectiveness of using the method of artificial intelligence in maintenance of ICT systems. Management systems. Production Engineering. 2019;**1**(27):40-45

[39] Kremer OS, Cunha MA, Silva PH. Implementation of a predictive maintenance system using unsupervised anomaly detection. In: Simpósio Brasileiro de Automação Inteligente-SBAI. 2021;**1**(1). DOI: 10.20906/sbai.v1i1.2783. ISSN: 2175-8905

[40] Achouch M, Dimitrova M, Ziane K, Sattarpanah Karganroudi S, Dhouib R, Ibrahim H, et al. On predictive maintenance in industry 4.0: Overview, models, and challenges. Applied Sciences. 2022;**12**(16):8081. DOI: 10.3390/app12168081

[41] Pech M, Vrchota J, Bednář J. Predictive maintenance and intelligent sensors in smart factory: Review. Sensors. 2021;**21**(4):1470. DOI: 10.3390/s21041470

[42] Faheem HB, Mahmoud AA, Hashem MD, Abd El Moez A, M. Modeling of pavement maintenance decisions using artificial intelligence based on maintenance unit. MEJ. Mansoura Engineering Journal. 2022;**47**(3):10-21

[43] Zhang M, Amaitik N, Wang Z, Xu Y, Maisuradze A, Peschl M, et al. Predictive maintenance for remanufacturing based on hybrid-driven remaining useful life prediction. Applied Sciences. 2022;**12**(7):3218. DOI: 10.3390/app12073218

[44] Rødseth H, Schjølberg P, Wabner M, Frieß U. Predictive maintenance for synchronizing maintenance planning with production. In: Wang K, Wang Y, Strandhagen J, Yu T, editors. Advanced Manufacturing and Automation VII 7. IWAMA 2017. Lecture Notes in Electrical Engineering. Singapore: Springer; 2018;**451**:439-446. DOI: 10.1007/978-981-10-5768-7_47

[45] Sang GM, Xu L, de Vrieze P, Bai Y, Pan F. Predictive maintenance in industry 4.0. Proceedings of the 10th International Conference on Information Systems and Technologies. June 2020:1-11

[46] Midha A, Maheshwari I, Ojha K, Gupta K, Deshpande SV. Bibliometric review of predictive maintenance using vibration analysis. Library Philosophy and Practice (e-journal). 2021:5790. ISSN 1522-0222

[47] Gulati K, Basandrai K, Tiwari S, Kamat P, Kumar S. Predictive Maintenance of Bearing Machinery Using Simulation-a Bibliometric Study. 2021

[48] Nalbantov G, Todorov D, Zografov N, Georgiev S, Bojilova N. Predictive maintenance tool for non-intrusive inspection systems. arXiv preprint arXiv:2103.01044. 2021

[49] Chen C, Liu Y, Wang S, Sun X, Cairano-Gilfedder CD, Titmus S, et al. Predictive maintenance using cox proportional hazard deep learning. Advanced Engineering Informatics. 2020;**44**:101054

[50] Bosman LB, Leon-Salas WD, Hutzel W, Soto EA. PV system predictive maintenance: Challenges, current approaches, and opportunities. Energies. 2020;**13**(6):1398. DOI: 10.3390/en13061398

[51] Cheng W, Feng H, Liang G. Design of IT infrastructure multicloud management platform based on hybrid cloud. Wireless Communications and Mobile Computing. 2022;**2022**. Article ID 9227948:12. DOI: 10.1155/2022/922794